Memoirs of the American Mathematical Society

Number 209

I. M. Sigal

Mathematical foundations of quantum scattering theory for multiparticle systems

Published by the

AMERICAN MATHEMATICAL SOCIETY

Providence, Rhode Island

VOLUME 16 · ISSUE 2 · NUMBER 209 (end of volume) · NOVEMBER 1978

ABSTRACT

This paper is devoted to scattering theory for multiparticle systems. The main result is a proof of the completeness of the scattering eigenfunctions for systems of an arbitrary but fixed number of particles.

Hitherto, theorems of this kind were known for two-particle ([1, 2]), three-particle ([3]) and four-particle [4] systems and for systems of any number of particles with finite potentials for which the Schrödinger operators of the subsystems have no point spectrum ([4]), with repulsive ([5]) and small ([6]) potentials. Furthermore, a proposition closely related to our main result (Theorem 3.5 of this paper) was stated without proof in [4].

A brief account of this paper was published in [7].

AMS(MOS) subject classifications (1970). Primary 47A40, 81A48; Secondary 46E35, 47A55, 47A70, 47F05.

Key words and phrases. N-body system, scattering theory, wave operators, behaviour of resolvent near the real axis.

Library of Congress Cataloging in Publication Data CIP

Sigal, Israel Michael, 1945-
Mathematical foundations of quantum scattering theory for multiparticle systems.

(Memoirs of the American Mathematical Society ; no. 209)
"Volume 16, issue 2."
Includes bibliographical references.
1. Scattering (Physics) 2. Quantum theory.
I. Title. II. Series: American Mathematical Society. Memoirs ; no. 209.
QA3.A57 no. 209 [QC794.6.S3] 510'.8s [539.7'54]
ISBN 0-8218-2209-8 78-10154

TABLE OF CONTENTS

MATHEMATICAL FOUNDATIONS OF QUANTUM SCATTERING THEORY FOR MULTIPARTICLE SYSTEMS*

1. INTRODUCTION

The evolution of a nonrelativisitic quantum system is described by the one-parameter group $\exp(-iHt)$, where H is the Schrödinger operator of the system. The task of scattering theory is to study the asymptotic behavior of the evolution operator $\exp(-iHt)$ for large times, $t \to \pm\infty$.

As was shown in [1,3,4], using the connection between the one-parameter group $\exp(-iHt)$ and the resolvent $R(z) = (H-z)^{-1}$, this corresponds to the description of the behaviour of $R(z)$ as z approaches $\sigma_c(H)$.

While the study of $\exp(-iHt)$ for large t's leads to the nonstationary formalism of scattering theory, the study of $R(z)$ near the real axis is related to the stationary one. The latter aims in the description of the asymptotic behaviour in the configuration space of the generalised eigenfunctions of H (solutions of the stationary Schrödinger equation). It was shown in [1-3] that certain information about $R(z)$ is sufficient for the complete justification of the time-dependent scattering theory as well.

It appeared that the resolvent $R(z)$ is a quite suitable object for investigation. It was shown in the cases of one(two)-body systems [1,2] and the three body systems [3] that its boundary values on $\sigma_c(H)$ can be obtained as solutions of an appropriate equation in an appropriate space. In the mentioned papers the principal mathematical problems of the scattering theory for the corresponding systems were solved.

Received by the editors June 30, 1975 and, in revised form July 21, 1976.

* Supported in part by U.S.-Israel Binational Science Foundation (B.S.F.), Jerusalem, Israel.

The further development in this direction came with [4], where the scattering theory for four-body systems with fast decreasing potentials and n-body single-channel systems with exponentially vanishing potentials was justified.

Other and somewhat simpler methods were developed on the basis of the Kato theory of relatively smooth operators ([39]): the mathematical justification of the scattering theory for n-body, single channel systems with repulsive [5] and small [6] potentials was given in the stationary approach.

In this paper we study the basic mathematical problems of scattering theory for multiparticle systems with short range potentials (i.e. potentials vanishing at infinity faster than $|x|^{-2}$). We are using the stationary approach or more specifically we are working in the general framework proposed in [3] for three particle systems. The key point of the paper is a theorem on the behavior of the resolvent of the Schrödinger operator for a system with an arbitrary but fixed number of particles, as the complex parameter approaches the continuous spectrum of the operator (Theorem 3.5). A similar theorem was stated without a proof in [4]. The theorem is then used to prove the existence and completeness of the scattering eigenfunctions, i.e., the generalized eigenfunctions of the operator H satisfying prescribed asymptotic conditions at infinity in configuration space. It will be shown that the scattering eigenfunctions are the kernels of wave operators. This implies a justification of the fundamental problem of quantum scattering theory for multiparticle systems. The unitary operator performing the spectral representation for H is constructed from the wave operators.

The restrictions on the potentials used in this work consist of direct estimates (2.11), (2.12) of smoothness and fall-off at infinity and indirect conditions A and B on the point spectrum of the Schrödinger operators of the subsystems and their properties at the threshold values, respectively. The direct restrictions provide, essentially, that the potentials fall off at infinity as $|x|^{-2-\varepsilon}$, $\varepsilon = \nu_o - \frac{1}{2} > 0$, and the indirect conditions require the absence of the bound states and quasibound states respectively in the continuous spectrum of the Schrödinger operators of the subsystems. It can be shown that the indirect conditions are satisfied for the rather large class of potentials (for the precise statement and discussion see Section 3).

The proof of Theorem 3.5 is based on Equation (4.10) for the resolvent of H, derived by F.A. Berezin [9] who used it for somewhat different purposes.

For the investigation of this equation, we have chosen the so-called momentum representation for the reasons that the kernels of the operators under investigation have a simpler form and the desired properties of the resolvent $(H-zI)^{-1}$ are described more easily.

In this work, we restrict ourselves to the consideration only of two-body interactions, but the generalization of the results to many body interactions is not difficult.

As another easily derived generalization, we mentioned the possibility of taking into consideration the symmetry properties of a system by carrying out the study in the invariant subspaces of fixed types of irreducible representations of a group of symmetry as was done, for example, in [10, 11] (see also the discussion of Condition A in Section 3, §3.2).

The article is divided into seven sections with three appendices in which most of the technical estimates are carried out, and one supplement where some frequently used theorems are collected. The central and most difficult section is Section 6. It is preceded by the outline of its content. Each section is divided into paragraphs. The statements (theorems, propositions and lemmas) are enumerated by the number of the paragraph in which they occur. We wish to turn the reader's attention to the specific enumeration of the formulas used in this article. Often an expression is numbered, not only by its ordinary number, but also by a value of the parameter, namely decomposition, which is contained in this expression. Thus the reference, for example, to equation (4.34b) means reference to equation (4.34) for the value of the decomposition, by which it is labelled, equal to b.

In conclusion, the author wishes to record that the main ideas of this paper originated in discussions with F.A. Berezin. The author is indebted to F.A. Berezin for his constant attention to the work and his suggestive advice. Thanks are also due to L.D. Faddeev and D.R. Yafaev, whose comments were instrumental in eliminating some serious inaccuracies, and to G.M. Zhislin for discussions on various questions touching upon the subject.

2. STATMENT OF THE PROBLEM. NOTATIONS

Let I be a finite subset of the set of natural numbers. We associate with I an operator in $L_2(R^{3 \cdot \sum_{i \in I} 1})$

$$(H_I f)(x) = - \sum_{i \in I} \frac{1}{2m_i} (\Delta_i f)(x) + \frac{1}{2} \sum_{\substack{i,j \in I \\ i \neq j}} V_{ij}(x_i - x_j) f(x)$$

$$= (H_o^I f)(x) + (V^I f)(x),$$

where $m_i > 0$, $x_i \in R^3$, and Δ_i is the Laplacian in the coordinate x_i. Under quite broad assumptions on $V_{ij}(x)$, the operator H_I is selfadjoint on its natural domain of definition. Following the accepted terminology of quantum mechanics, we shall call the pair (I, H_I) a multiparticle system (or simply system), and H_I will be called the Schrödinger operator of the system. Henceforth we shall apply the term system to the set I, since throughout the sequel V_{ij} will be fixed operators from a certain class to be described below, and so the operator H_I is uniquely determined by I.

If $I = \{1, \ldots, n\}$, we set

$$H_I = H_n = H_o^{(n)} + V^{(n)} .$$

The fundamental problem of the quantum scattering theory for systems of n particles is to determine the asymptotic behavior as $t \to \infty$ of the solutions of the Schrödinger equation

$$i \frac{\partial \psi}{\partial t} = H_n \psi , \qquad (2.1)$$

on the assumption that their asymptotic behavior as $t \to -\infty$ is known.

The asymptotic behavior of the solutions of Eq. (1.1) as $t \to -\infty$ should be chosen subject to physical considerations. To describe it let

$\underline{a} = \{C_i\}$ be a collection of nonempty and nonintersecting subsets of $\{1,\ldots,n\}$ such that $\cup C_i = \{1,\ldots,n\}$. We shall call any such collection a partition, and the subsets C_i will be called subsystems. We denote the number of subsystems in a partition $\underline{a}$ by $k(a)$, and introduce the following notation:

$$A_s = \{a\colon k(a) = s\},\ A' = \bigcup_2^n A_s,\ A = \bigcup_1^n A_s.$$

With each partition $\underline{a}$ we associate an operator

$$H_a^{(n)} = H_o^{(n)} + \sum_{C_i \varepsilon a} V^{C_i},\ H_a^{(n)} = H_n (a \in A_1),\ H_a^{(n)} = H_o^{(n)}\ (a \in A_n)$$

The operators $H_a^{(n)}$, $a \in A$, commute with the total momentum operators $P_{C_k} = \sum_{j \varepsilon C_k} \frac{1}{i} \nabla_j$ of the subsystems in the partition $\underline{a}$.

Let $\tilde{\psi}^{a,m}$ denote the generalized eigenfunctions of the operators $H_a^{(n)}$, annihilated by the operators P_{C_k}, $C_k \in a$, and square integrable with respect to a certain set of $n - k(a)$ variables from the set $\{x_1,\ldots,x_n\}$; these variables are so chosen that they include $k(C_i) - 1$ variables x_j, $j \in C_i$ $(i = 1,2,\ldots)$, where $k(C_i)$ is the number of particles (indices) in the subsystem C_i. In the rest of this section we shall confine attention to partitions $\underline{a}$ from $\bigcup_1^{n-1} A_s$ for which the operators $H_a^{(n)}$ have generalized eigenfunctions of the type stipulated.

Define $\tilde{\mathscr{U}}_{a,m}$, $m = 1,\ldots,m(a)$, $a \in A$, to be the space of all functions

$$\tilde{\psi}^{a,m}(x_1,\ldots,x_n) f(y_1,\ldots,y_{k(a)}),$$

where $y_i = (\sum_{j \in C_i} m_j)^{-1} \sum_{j \in C_i} m_j x_j$, $f \in L_2(R^{3k(a)})$.

One stipulates that the "asymptotes" of the solutions of Eq. (2.1) as $t \to -\infty$ be solutions of the equations

$$i\frac{\partial\psi}{\partial t} = H_a^{(n)}\psi \tag{2.2}$$

with initial conditions in the spaces $\tilde{\mathscr{A}}_{a,m}$ $(a \in A')$.

Physics considerations suggest the boundary conditions as $t \to \infty$ for solutions of Eq. (2.1) having a prescribed asymptotic behavior as $t \to -\infty$. In fact, their asymptotes must be superpositions of solutions of Eqs. (2.2) with initial conditions in $\tilde{\mathscr{A}}_{a,m}$ for various partitions $\underline{a}$.

We let $\psi_{a,m}(t)$ denote the solutions of Eq. (2.2) with initial conditions in $\tilde{\mathscr{A}}_{a,m}$.

Our original problem now splits up naturally into two problems:

I. Find solutions $\psi^-(t)$ of Eq. (2.1) possessing a prescribed asymptotic behavior $\psi_{a,m}(t)$ as $t \to -\infty$.

II. Find solutions $\psi^+(t)$ of Eq. (2.1) posssessing a prescribed asymptotic behavior $\psi_{a,m}(t)$ as $t \to +\infty$.

The original problem will ultimately be solved by expanding solutions of Eq. (2.1) of the first type in terms of solutions of the same equations of the second type.

Since the states of a quantum system are identified (by extrapolation) with the solutions of Eq. (2.1), the question whether any solution of Eq. (2.1) may be expanded in terms of solution of the first (or second) type rises in accordance with the physical picture of the elementary processes. Thus, in addition to the two problems just formulated, we have a third:

III. Prove that the solutions of Eq. (2.1) described in problem I (II) constitute, in some sense, a complete system of solutions of Eq. (2.1).

We now rephrase the three problems in a more detailed form.

Let us write the solutions of Eqs. (2.1) and (2.2) as

$$\psi(t) = e^{-iH_n t}\psi, \quad \psi_{a,m}(t) = e^{-iH_{a,m}^{(n)}t}\psi_{a,m} \tag{2.3}$$

$$(\psi \in L_2(R^{3n}), \quad \psi_{a,m} \in \tilde{\mathscr{A}}_{a,m}),$$

where $H_{a,m}^{(n)}$ denotes the restriction of $H_a^{(n)}$ to $\tilde{\mathcal{A}}_{a,m}$. In view of the probability interpretation of the Schrödinger equation, we understand $\psi_{a,m}(t)$ to be an asymptote of $\psi^{\pm}(t)$ in the sense of the L_2-topology:

$$|\psi^{+}(t) - \psi_{a,m}(t)| \to 0 \quad \text{as} \quad t \to \pm\infty , \tag{2.4}$$

where $|\cdot|$ (without subscript) denotes the L_2-norm. Inserting (2.3) into (2.4) and using the unitarity of the operator $\exp(-itH_n)$, we obtain

$$|\psi^{\pm} - e^{iH_n t} e^{-iH_{a,m}^{(n)} t} \psi_{a,m}| \to 0 \quad \text{as} \quad t \to \pm\infty .$$

Hence the following mathematical formulation of problems I and II:

Prove the existence of

$$\tilde{U}_{a,m}^{(\pm)} = \underset{t\to\pm\infty}{\text{s-lim}}\; e^{itH_n} e^{-itH_{a,m}^{(n)}} , \tag{2.5}$$

where s-lim denotes the strong limit in $\tilde{\mathcal{A}}_{a,m}$.

And problem III is formulated thus:

Prove that the operators $\tilde{U}_{a,m}^{(\pm)}$ are complete, that is to say,

$$\sum_{a,m} \tilde{U}_{a,m}^{(\pm)} \tilde{U}_{a,m}^{(\pm)*} = E.$$

Any pair (a,m) is usually called a reaction (scattering) channel, $H_{a,m}^{(n)}$ are the channel Hamiltonians and $\tilde{U}_{a,m}^{(\pm)}$ the wave operators corresponding to the channel (a,m). In terms of wave operators, the original problem -- to determine the asymptotic behavior as $t \to -\infty$ -- may be solved as follows. It follows from (2.5), (2.6) that any solution $\psi(t)$ of Eq. (2.1) may be expanded in solutions whose asymptotes as $t \to -\infty$ have the form $e^{-i\tilde{H}_{a,m}^{(n)} t}\phi$, $\phi \in \tilde{\mathcal{A}}_{a,m}$:

$$\psi(t) = \sum_{a,m} e^{-iH_n t} \tilde{U}_{a,m}^{(-)} \tilde{U}_{a,m}^{(-)*} \psi(0) ,$$

$$e^{-iH_n t}\,\tilde{U}^{(-)}_{a,m}\,\tilde{U}^{(-)*}_{a,m}\,\psi(0) \sim e^{-iH^{(n)}_{a,m}t}\,\tilde{U}^{(-)*}_{a,m}\,\psi(0) \qquad (t \to -\infty).$$

The asymptotic behavior as $t \to +\infty$ of a solution of Eq. (2.1) whose asymptote at $t \to -\infty$ is $e^{-iH^{(n)}_{a,m}t}\psi_o$, $\psi_o \in \tilde{\mathscr{A}}_{a,m}$, is given by

$$\sum_{b,k} e^{-iH^{(n)}_b t}\,\tilde{U}^{(+)*}_{b,k}\,\tilde{U}^{(-)}_{a,m}\,\psi_o .$$

Indeed, consider the following solution of Eq. (2.1):

$$e^{-iH_n t}\,\hat{\tilde{U}}^{(-)}_{a,m}\,\psi_o .$$

We have

$$\left|e^{-iH_n t}\,\tilde{U}^{(-)}_{a,m}\,\psi_o - e^{-iH^{(n)}_{a,m}t}\,\psi_o\right| = \left|\tilde{U}^{(-)}_{a,m}\,\psi_o - e^{+iH_n t}\,e^{-iH^{(n)}_{a,m}t}\,\psi_o\right| \to 0$$

as $t \to -\infty$, and

$$\left|e^{-iH_n t}\,\tilde{U}^{(-)}_{a,m}\,\psi_o - \sum_{b,k} e^{-iH^{(n)}_{b,k}t}\,\tilde{U}^{(+)*}_{b,k}\,\tilde{U}^{(-)}_{a,m}\,\psi_o\right|$$

$$= \left|\tilde{U}^{(-)}_{a,m}\,\psi_o - \sum_{b,k} e^{+iH_n t}\,e^{-iH^{(n)}_{b,k}t}\,\tilde{U}^{(+)*}_{b,k}\,\tilde{U}^{(-)}_{a,m}\,\psi_o\right|$$

$$\to \left|\tilde{U}^{(-)}_{a,m}\,\psi_o - \sum_{b,k} \tilde{U}^{(+)}_{b,k}\,\tilde{U}^{(+)*}_{b,k}\,\tilde{U}^{(+)}_{a,m}\,\psi_o\right| = 0 \quad \text{as} \quad t \to -\infty.$$

The operator S defined in the space

$$\tilde{\mathscr{A}} = \sum_{a,m} + \tilde{\mathscr{A}}_{a,m}$$

by the matrix

$$\tilde{S}_{a,m;b,k} = \tilde{U}^{(+)*}_{a,m}\,\tilde{U}^{(-)}_{b,k}$$

is known as the scattering operator (see [12]).

In order to obtain the stationary form of the wave operators, one writes (2.5) as the Abel limit

$$s - \lim_{t\to\pm\infty} e^{iH_n t} e^{-iH_{a,m}^{(n)}t} = s - \lim_{\varepsilon\downarrow o} \pm \varepsilon \int_o^{\pm\infty} e^{\pm\varepsilon t} e^{iH_n t} e^{-H_{a,m}^{(n)}t} dt$$

(the right side of this equality always exists when the left side exists; the converse is not true) and use the representation of $\exp(-iH_{a,m}^{(n)}t)$ through the resolution of the identity, $E_{a,m}^{(n)}$, for $H_{a,m}^{(n)}$:

$$\tilde{U}_{a,m}^{(\pm)} = s - \lim_{\varepsilon\downarrow o} \pm \varepsilon \int e^{\varepsilon t} e^{iH_n t} e^{-i\lambda t} dE_{a,m}^{(n)}(\lambda)\, dt =$$

$$= s - \lim_{\varepsilon\downarrow o} \mp i\varepsilon \int R_n(\lambda \mp i\varepsilon)\, dE_{a,m}^{(n)}(\lambda) \tag{2.7}$$

In order to pass to the last expression we changed the order of integration, which can easily be justified [13], and used the well-known formulae connecting the evolution operator and the resolvent of a self-adjoint operator

$$(A - zI)^{-1} = i \int_o^{\pm\infty} e^{-izt} e^{iAt} dt\ , \qquad z \in C^{\pm}$$

The Schrödinger operator of a system of n particles with pair interaction has the following form in the momentum representation (we are retaining the same notation for the operator as in the coordinate representation):

$$H_n = H_o^{(n)} + V^{(n)}, \quad V^{(n)} = \frac{1}{2} \sum_{\substack{i,j=1 \\ i\neq j}}^{n} V_{ij}^{(n)}\ , \tag{2.8}$$

$$(H_o^{(n)} f)(p) = \sum_{i=1}^{n} \frac{p_i^2}{2m_i} f(p), \tag{2.9}$$

$$(V_{ij}^{(n)} f)(p) = \int v_{ij}(p_i - q_i)\delta(p_i + p_j - q_i - q_j) \prod_{k\neq i,j} \delta(p_k - q_k) f(q) d^n q, \tag{2.10}$$

where $m_i > 0$, $v_{ij}(-k) = \overline{v}_{ij}(k)$, $p_i \in R^3$, $p = (p_1,\ldots,p_n)$. Throughout this paper we shall assume that $v_{ij}(k)$ satisfies the following conditions [3]:

$$|v_{ij}(k)| < C(1 + |k|)^{-\eta_o}, \tag{2.11}$$

$$|v_{ij}(k + h) - v_{ij}(k)| < C(1 + |k|)^{-\eta_o}|h|^{\nu_o}, \ \eta_o > 3/2, \ \nu_o > 1/2. \tag{2.12}$$

If condition (2.11) is satisfied, the oprator H_n is defined on $S(R^{3n})$ and the subordination inequality holds [14]:

$$|V^{(n)}f| \leq C(\rho^{\alpha}|H_o^{(n)}f| + \rho^{-\frac{3n}{2}}|f|), \ f \in S(R^{3n}), \ \alpha > 0, \tag{2.13}$$

where $\rho > 0$ is an arbitrary number. It follows from (2.13) (see [14]) that H_n has a unique selfadjoint extension.

In the momentum space R^{3n} of the system, we define (see [10]) the subspace R of relative movement:

$$R = \{p: p = (p_1,\ldots,p_n) \in R^{3n}, \ p_1 + \ldots + p_n = 0\},$$

and the subspace R_c of center-of-mass (CM) movement of the system:

$$R_c = \{p: p = (p_1,\ldots,p_n) \in R^{3n}, \ m_1^{-1}p_1 = \ldots = m_n^{-1}p_n\}.$$

We now define in R^{3n} a scalar product

$$(p,\tilde{p})' = \sum_{i=1}^{n} \frac{1}{m_i}(p_i, \tilde{p}_i), \tag{2.14}$$

where $(p_i,\tilde{p}_i)$ is the standard scalar product in R^3. Then it is readily seen that the subspaces R and R_c are mutually orthogonal in the sense of this scalar product, and moreover

$$R \oplus R_c = R^{3n}. \tag{2.15}$$

It follows from (2.15) that

$$L_2(R^{3n}) = L_2(R) \otimes L_2(R_c).$$

This decomposition induces a decomposition of the operator H_n:

$$H_n = H \otimes E_c + E \otimes H_c, \tag{2.16}$$

where E and E_c are the identity operators in $L_2(R)$ and $L_2(R_c)$, respectively, H is defined in $L_2(R)$, H_c is defined in $L_2(R_c)$ and has the form

$$(H_c f)(p) = (2 \sum_{i=1}^{n} m_i)^{-1} (\sum_{i=1}^{n} p_i)^2 f(p).$$

To develop an analytical expression for H, we proceed as follows. The image of a function $f(p) = f_o(p)\delta(\sum p_i)$, $f_o \in S(R)$, under any operator $V_{ij}^{(n)}$ is a function of the form $\phi(p) = \phi_o(p)\delta(\sum p_i)$, $\phi_o \in L_2(R)$. Thus the operators $V_{ij}^{(n)}$ induce operators in $L_2(R)$, which we denote by V_{ij}. Set

$$V = \sum_{\substack{i,j=1 \\ i<j}}^{n} V_{ij}.$$

Define an operator H_o by

$$(H_o f)(p) = \sum_{i=1}^{n} \frac{p_i^2}{2m_i} f(p), \quad p \in R, \ f \in S(R).$$

Then H may be expressed as

$$H = H_o + V.$$

The subordination inequality (2.17) for H_n implies (see [10]) a subordination inequality for the operator H:

$$|Vf| \leq C(\rho^{\alpha}|H_o f| + \rho^{-\frac{3n}{2}}|f|), \ f \in S(R), \ \alpha > 0, \tag{2.17}$$

where ρ is an arbitrary positive number. Thanks to this inequality, H has a unique selfadjoint extension [14], for which we retain the old notation H.

We now proceed to the description of compound systems, i.e., systems derived from the original system when one neglects the interaction between certain of its subsystems. Define the subspace R^a of relative movement of the subsystem in a partition $\underline{a}$ (each subsystem relative to its center of mass):

$$R^a = \{p: p = (p_1,\dots,p_n) \in R, \sum_{j\varepsilon C_i} p_j = 0, C_i \in a\},$$

and the subspace of relative movement of the centers of mass of these subsystems:

$$R_a = \{p: p = (p_1,\dots,p_n) \in R, p_i = m_i(\sum_{j\varepsilon C_k} m_j)^{-1} P_k, P_k \in R^3, i \in C_k, C_k \in a\}.$$

As before, we have the following relationships, in the sense of the scalar product (2.14):

$$R^a \perp R_a, \quad R^a \oplus R_a = R, \quad L_2(R) = L_2(R^a) \otimes L_2(R_a). \tag{2.18}$$

Recall that $H_a^{(n)}$ denotes the operator obtained from H_n by deleting interaction operators between particles in different subsystems of the partition $\underline{a}$. We can express this operator as

$$H_a^{(n)} = H_a \otimes E_c + E \otimes H_c, \quad H_a = H^a \otimes E_a + E^a \otimes T_a, \tag{2.19}$$

where E^a and E_a are the identity operators in $L_2(R^a)$ and $L_2(R_a)$, respectively; H^a is the operator of movement of the subsystems $C_i \in a$, each relative to its center of mass, defined in $L_2(R^a)$; T_a is the operator of relative movement of the centers of mass of the subsystems $C_i \in a$, defined in $L_2(R_a)$ by

$$(T_a f)(p) = \sum_{j=1}^{k(a)} (2\sum_{i\varepsilon C_j} m_i)^{-1} (\sum_{i\varepsilon C_j} p_i)^2 f(p).$$

The operators H_a and H^a satisfy subordination inequalities of type (2.17), and they have unique selfadjoint extensions.

Let

$$\bar{A} = \{a,\ a \in A\ ,\ H^a \text{ has a point spectrum}\} \cup A_n .$$

Let $\psi^{a,m}$ and $\lambda^{a,m}$ $(m = 1,\ldots,m(a){:}\ m(a) = 1,2,\ldots)$ denote the eigenfunctions and the corresponding eigenvalues of H^a, $a \in \bar{A}$ (the $\lambda^{a,m}$ are not necessarily distinct). If $a \in A_n$ we set $m(a) = 1$, $\lambda^a = 0$, $\psi^a \equiv 1$. $\{\lambda^{a,m},\ m = 1,2,\ldots,a \in \bar{A}\}$ is called threshold set for the Schrödinger operator H.

Define the spaces

$$\mathcal{A}_{a,m} = \psi^{a,m}(p^a) \otimes L_2(R_a) .$$

Recalling the spaces $_{a,m}$ of sec. 1, we note that

$$F\tilde{\mathcal{A}}_{a,m} = \mathcal{A}_{a,m} \otimes L_2(R_c) .$$

This decomposition induces a representation of the operators $\tilde{U}^{(\pm)}_{a,m}$:

$$F\tilde{U}^{(\pm)}_{a,m}F^* = U^{(\pm)}_{a,m} \otimes E_c ,$$

where $U^{(\pm)}_{a,m}$ are the operators defined in $\mathcal{A}_{a,m}$ by

$$U^{(\pm)}_{a,m} = \underset{t\to\pm\infty}{\text{s-lim}}\ e^{iHt}\, e^{-iH_{a,m}t} ,$$

where $H_{a,m}$ is the restriction of H_a to $\mathcal{A}_{a,m}$.

In terms of the operators $U^{(\pm)}_{a,m}$, the completeness relation (2.6) becomes

$$\sum_{a,m} U^{(\pm)}_{a,m}\, U^{(\pm)*}_{a,m} = E. \tag{2.20}$$

The existence of the wave operators $U^{(+)}_{a,m}$ implies their isometry and orthogonality ([15]).

$$U^{(\pm)*}_{a,m} U^{(\pm)}_{b,k} = \delta_{a,b}\, \delta_{m,k}$$

and their intertwining property

$$HU^{(\pm)}_{a,m} = U^{(\pm)}_{a,m} H_{a,m} .$$

Thus the fundamental problem of quantum scattering theory reduces to proving the existence and completeness (i.e., the relation (2.20)) of the wave operators $U^{(\pm)}_{a,m}$, $m = 1,\dots,m(a)$, $a \in \bar{A}$.

The first part of this problem -- the existence of the operators $U^{(\pm)}_{a,m}$ -- was solved in [16, 17], where an easily checked existence criterion was established. The second part -- the completeness proof -- has been solved only in a few special cases (two-particle [1, 2], three-particle [3] and four-particle [4] systems, and some particular cases of n-particle systems [4-6]; see Introduction). This paper presents a proof of (2.20) for systems of an arbitrary but fixed number of particles, subject to certain restrictions on the potentials which are n-particle generalizations of the assumptions adopted in [3].

We end this section with a few definitions and notational conventions.

If b is a partition obtained by breaking up certain subsystems of a, we shall say that b is contained in the partition a, writing $b \subset a$. The smallest partition containing two partitions a,b will be denoted by $a \cup b$, i.e., $a \cup b = \sup(a,b)$. The largest partition contained in both a and b will be denoted by $a \cap b$: $a \cap b = \inf(a,b)$. Throughout this paper, the symbol $\subset$ will always stand for proper inclusion, improper inclusion being denoted by $\subseteq$.

We introduce the notation

$$R^a_b = R^a \cap R_b = R^a \ominus R^b , \quad b \subset a.$$

Apart from the decompositions (2.16) and (2.19), an operator H_b admits certain decompositions following from (2.18) $(a \supseteq b)$:

$$H_b = H_b^a \otimes E_a + E^a \otimes T_a,$$

where H_b^a is an operator defined in $L_2(R^a)$, or, in more general form,

$$H_b^a = H_b^d \otimes E_d^a + E^d \otimes T_d^a, \quad b \subset d \subseteq a, \tag{2.21}$$

where E_d^a is the identity operator in $L_2(R_d^a)$ and

$$(T_d^a f)(p_d^a) = \frac{1}{2}(p_d^a, p_d^a)' f(p_d^a), f(p_d^a) \in L_2(R_d^a).$$

To derive an analytical expression for H_b^a, $b \subseteq a$, we proceed as follows. We insert a function of the form

$$f(p) = f_a(p^a) \prod_{C_i \varepsilon a} \delta(p_{C_i}), \quad f_a \in S(R^a),$$

into the expression $H_b^{(n)} f$. Then the operator generated by $H_b^{(n)}$ in $L_2(R^a)$ is precisely H_b^a. Similarly, the operators $V_\alpha^{(n)}$ generate operators V_α^a in $L_2(R^a)$. The operator H_b^a may be expressed in terms of V_α^a as follows:

$$(H_b^a f)(p^a) = \frac{1}{2}(p^a, p^a)' f(p^a) + \sum_{u_\alpha \in b} (V_\alpha^a f)(p^a), \quad u_\alpha = \{(\alpha) \prod_{k \notin \alpha} (k)\}$$

We have

$$H_a^a = H^a \text{ and } H_b^a = H_b \text{ for } a \in A_1.$$

Let $R_{b,a}(z)$ denote the resolvent of the operator H_b^a, $R^a(z) = R_{a,a}(z)$. Formula (2.14) implies a relation between operators $R_{b,a}(z)$ for different a, $a \supseteq b$:

$$R_{b,a}(z) = R_{b,d}(z - \frac{1}{2}(p_d^a, p_d^a)') \otimes E_d^a, \quad b \subseteq d \subset a. \tag{2.22}$$

For $f \in L_2(R^a)$ we denote by $f(p_d^a)$ the function f which is considered as a vector-function from R_d^a into $L_2(R^d)$. Then the right side of (2.15) means that

$$(R_{b,d}(z - \tfrac{1}{2}(p_d^a, p_d^a)') \otimes E_d^a)f = R_{b,d}(z - \tfrac{1}{2}(p_d^a, p_d^a)')f(p_d^a), \quad f \in L_2(R^a) \tag{2.22'}$$

If $b \in A_n$, we shall employ the notation

$$H_b^a = H_o^a,\ R_{b,a}(z) = R_{o,a}(z),$$

which originates from the universally accepted notation.

Let $P^{a,m}$ denote the eigenprojection corresponding to the eigenvalue $\lambda^{a,m}$ of H^a, $a \in \bar{A}$. The discrete part of $R^a(z)$ will be denoted by $P^a(z)$:

$$P^a(z) = \sum_{m=1}^{m(a)} (\lambda^{a,m} - z)^{-1} P^{a,m} \quad \text{for } a \in \bar{A}, \quad P^a(z) = 0 \quad \text{for } a \notin \bar{A}$$

The same thing for $R_{b,a}(z)$ is

$$P_{b,a}(z)f = P^b(z - \tau(p_b^a))f(p_b^a), \quad f \in L_2(R)$$

When $a \in A_1$ we will omit this index:

$$P_b(z) = P_{b,a}(z) \quad (a \in A_1).$$

Points of the spaces R, R^a, R_a and R_b^a will be denoted, respectively, by p,q; p^a,q^a; p_a,q_a and p_b^a,q_b^a. In accordance with the decomposition (2.18), variables in R may be expressed as

$$p = (p^a, p_a), \quad p^a \in R^a, \quad p_a \in R_a.$$

This equality means that the components $(p_1,\dots,p_n)$, $(p_1^o,\dots,p_n^o)$ and $(p_1^c,\dots,p_n^c)$ of the vectors p, p^a and p_a satisfy the relations

$$p_j^o = p_j - m_j\Big(\sum_{s\in C_i} m_s\Big)^{-1} \sum_{s\in C_i} p_s, \quad p_j^c = m_j\Big(\sum_{s\in C_i} m_s\Big)^{-1} \sum_{s\in C_i} p_s, \quad j \in C_i,$$

$$p_j = p_j^o + p_j^c, \qquad j \in C_i.$$

The volume elements in the spaces R, R^a and R_a will be denoted, respectively, by dp, dq; dp^a, dq^a and dp_a, dq_a.

Any integral sign with no indication of the domain of integration implies integration over the entire domain of definition of the variables. For example,

$$\int f(p)\,dp = \int_R f(p)\,dp.$$

If $k \in R^3$, the length (norm) of k will be denoted by $|k|$, and its angular coordinates in a spherical coordinate system by $\omega(k)$.

The symbol $\delta(p_a - q_a)$ will denote the generalized kernel of the identity operator E_a in $L_2(R_a)$, $a \in A$.

We shall also use the notation

$$\tau(p) = \frac{1}{2}(p, p)', \quad p \in R;$$

Note that

$$\tau(p_b^a) = \tau(p_b^d) + \tau(p_d^a), \quad b \subset d \subset a; \tag{2.23}$$

We define estimating functions*) in terms of which we will describe a fall-off of the functions at infinity. Let I be a set of k natural numbers,

$$\Delta = \{\delta = (\delta_i \ldots \delta_{k-1}): \delta_i \subset I,\ \delta_i \neq \delta_j,\ \delta_i \cap \delta_j \in \{\phi, \delta_i, \delta_j\}\ (i,j = 1 \ldots k-1),$$

$$\bigcup_1^{\ell} \delta_{i_t} \in \{\delta_i \in \delta,\ i \neq i_1, \ldots, i_\ell\}\ \forall i_1, \ldots, i_\ell \leqq k-1,\ i_s \neq i_t\}$$

The following proposition is obvious:

LEMMA. The transformation of variables $q^{\delta_j} = \sum_{i \in \delta_j} p_i$, $\delta_j \in \delta$, $q^I = \sum_{i \in I} p_i$ is nonsingular.

Estimating functions are defined as follows:

---

*) These estimating functions were introduced in [18].

$$N(p_j,\ j \in I,\ \eta) = \sum_{\delta \in \Delta} \prod_{\delta_i \in \delta} (1 + |\sum_{j \in \delta_i} p_j|)^{-\eta}$$

and for $b = \{C_k'\} \subset a = \{C_k\}$

$$N_b^a(p_b;\ \eta) = \prod_{C_i \in a} N(p_{C_k'},\ C_k' \subseteq C_i;\ \eta)\ ,\quad M_b^a(\eta) = \{N_b^a(\eta)\}^{-1}$$

We use the notations

$$\Delta(h)f = f(\xi+h) - f(\xi),\ \xi, h \in R^m,\quad \Delta^r(h) = |h|^{-r}\Delta(h),$$

$$\Delta_\xi(h)f = f(\xi+h,\eta) - f(\xi,\eta),\quad \Delta_\xi^r(h) = |h|^{-r}\Delta_\xi(h)\ .$$

We define spaces**)

$$H_{s,\eta}(R_b^a) = \{f\colon M_b^a(\eta)\ f \in L_\infty(R_b^a),\ |M_b^a(\eta)\Delta^s(h)f|_{L_\infty(R_b^a)} \leq C\}$$

with the norm

$$|f|_{H_s(R_b^a)} = |M_b^a(\eta)f|_{L_\infty(R_b^a)} + \sup_h |M_b^a(\eta)\Delta^s(h)f|_{L_\infty(R_b^a)}$$

Note that by (2.11), (2.12) $v_\alpha \in H_{\nu_o,\eta_o}(R^{u_\alpha})$, where

$$u_{ij} = \{(ij) \prod_{k \neq i,j} (k)\} = \{(ij)(1)\ldots(i-1)(i+1)\ldots(j-1)(j+1)\ldots(n)\} \in A_{n-1} \tag{2.33}$$

Norm in $H_{\nu_o,\eta_o}(R^{u_\alpha})$ will be denoted simply by $|\cdot|_o$:

$$|v|_o = |(1+|k|)^{\eta_o} v(k)|_{L_\infty(R^3)} + \sup_h |(1+|k|)^{\eta_o} |\Delta^{\nu_o}(h)v(k)|_{L_\infty(R^3)}\ .$$

We remark that (2.33) defines one-to-one correspondence between A_{n-1} and the set $\{(ij)\colon i,j = 1,\ldots,n\}$.

**) These spaces are n-particle generalizations of those used by L.D. Faddeev [4] in the three body problem. The similar spaces were applied to a study of the n-particle scattering problem by K. Hepp [5]. We use these spaces only to formulate our main result and for an investigation of the resolvent outside of the continuous spectrum.

Let Π^a denote the complex plane split along the real axis from $\mu^a = \min_{b\epsilon a} \inf H_b$ to ∞; Π_1 will denote the domain in the complex plane obtained from Π^a, $a \in A_1$, by cutting out the following points together with certain neighbourhoods thereof: the eigenvalues of the operator H on the half-line $[\mu, +\infty)$, $\mu = \mu^a$ ($a \in A_1$), and the points $\lambda^{a,m}$, $a \in \bar{A} \cap A_2$. We set $\Pi_1^a = \Pi^a$ if $a \in A'$ and $\Pi_1^a = \Pi_1$ if $a \in A_1$.

When we write that a vector-function, $f(z)$, $z \in \Pi^a$, satisfies some estimates, we mean that the limits of $f(\lambda + i\varepsilon)$, $\lambda \in [\mu^a, +\infty)$, as $\varepsilon \to \pm 0$, exist, at least, in the weakest topology of the spaces of values of $f(z)$ and the estimates are valid for $f(z)$, $z \in C \setminus [\mu^a, +\infty)$ and $f(\lambda \pm i0)$, $\lambda \in [\mu^a, +\infty)$.

Furthermore, as usual

$$\theta(x) = \begin{cases} 1, & x > 0 \\ 0, & x \lesssim 0 \end{cases}$$

Later we shall have occasion to deal with products of non-commuting factors, indexed by partitions with the same number of subsystems and arranged in an arbitrary but previously selected and fixed order. To fix ideas, therefore, we shall number partitions containing the same number of subsystems as follows: with any partition $a = \{C_1, \ldots, C_s\}$ we associate the permutation $g_a = (C_1) \ldots (C_s)$, which may be written

$$g_a = \begin{matrix} 1 \ldots n \\ i_1 \ldots i_n \end{matrix}$$

For example, $a = \{\{1,2,3\}, \{4,5\}\} \to g_a = (123)(45) = \begin{matrix} 12345 \\ 23154 \end{matrix}$. We now number the partitions $\underline{a}$ in accordance with the sequences $(i_1 \ldots i_n)$, assigning a higher index to a partition corresponding to a sequence occurring "later" in the lexicographical order: $a_1 \to i_1 \ldots i_n = 12\ldots n$, $a_2 \to i_1 \ldots i_n = 12\ldots n\ n-1$, and so on. Then a product

$$\prod_{a\varepsilon A_s} T_a = \prod_{i,a_i\varepsilon A_s} T_{a_i}$$

of noncommuting factors T_{a_i} will be understood as follows:

$$\prod_{a\varepsilon A_s} T_a = \prod_{i,a_i\varepsilon A_s} T_{a_i} = T_{a_N} T_{a_{N-1}} T_{a_{N-2}} \ldots T_{a_1} .$$

The product of the same factors in the reverse order will be denoted by

$\prod'_{a\varepsilon\ s} T_a = \prod'_{i,a_i\varepsilon\ s} T_{a_i}$:

$$\prod'_{a\varepsilon A_s} T_a = \prod'_{i,a_i\varepsilon A_s} T_{a_i} = T_{a_1} \ldots T_{a_{N-1}} T_{a_N} .$$

As usual, $\mathscr{L}(X,Y)$ will denote the Banach space of bounded operators from a Banach space X to a Banach space Y. The norm in this space will be denoted by $|\cdot|_{X\to Y}$.

Following the accepted terminology, we shall say that a topological space X is compactly embedded in a topological space Y if $X \subset Y$ and every bounded set in X is compact in Y.

Finally, throughout the paper, the letter C without subscripts will be used to denote various constants; we shall distinguish between constants by means of subscripts only within each individual chain of inequalities.

3. MAIN RESULTS. SCHEME OF INVESTIGATION

3.1 We begin with the discussion of the indirect conditions which we shall impose upon the operator H, i.e., on the functions $v_{ij}(k)$ in addition to direct condition (2.11) and (2.12).

Condition A: The functions $v_{ij}(k)$ $(i,j = 1,\dots,n)$ have the property that the operators H^a, $a \in A'$, have no point spectrum imbedded into the continuous one.

We remind that H^a, $a \in A'$, is the Schrodinger operator for the compound system corresponding to the partition $\underline{a}$, i.e., the system composed from non-interacting subsystems $C_i \in a$ each in its centre-of-mass frame.

The second indirect restriction on H, Condition B, is formulated in Section 6, §6.36. In order to explain it here, we need to introduce the notion of quasibound state. This notion is based on the following result obtained in Sections 4 and 6:

For each $a \in A$ there exist an operator $L^a(\lambda + i0)$ and a Banach space B^a such that $L^a(\lambda + i0)$ is bounded in B^a and has a purely discrete spectrum, and

$$f \in B^a: \; f + L^a(\lambda + i0)f = 0 \;\Leftrightarrow\; f \in D(H^a) \; : \; (H^a - \lambda E^a)f = 0$$

(f is a bound state of H^a) if $\lambda \neq \lambda^{b,m}$, $m = 1,\dots,m(b)$, $b \in \bar{A} \cap A_{k(a)+1}$. The same statement with the same Banach space, B^a, is true also for $\lambda - i0$.

We say that H^a has a quasibound state at a value $\nu \in \sigma_p(H^b)$, $b \subset a$, $b \in \bar{A} \cap A_{k(a)+1}\}$ if one of the equations

$$f + L^a(\nu \pm i0)f = 0$$

(or both) has a nontrivial solution in B^a. One of the possible motivations for this definition is that a quasibound state is a generalized eigenfunction of H^a vanishing at infinity and for a suitable small variation of the potentials it becomes a bound state of Hamiltonian H^a [19].

Condition B requires that the operators H^a, $a \in A'$, have no quasi-bound states.

3.2 There are two cases when one has complete information about the point spectrum of the Schrödinger operator H:

H does not have any eigenvalue if

(I) the potentials V_{ij}'s are sufficiently small [6, also see there the formulation of restrictions on the potentials]:

or

(II) the potential $V = \frac{1}{2} \sum_{i \neq j} V_{ij}$ is repulsive [20]. The last means that beside some regularity condition like

(1) $v(x) = \frac{1}{2} \sum_{i \neq j} v_{ij}(x_i - x_j) \in Q_\alpha(R^{(0)})$ for some $\alpha > 0$, i.e.

$$\int_{|x-y|_1 \leq 1} |v(y)|\,|x-y|_1^{4-3(n-1)-\alpha} dy \text{ is uniformly bounded for } x \in R^{(0)};$$

(2) for any $x \in R^{(0)}$, $x \neq 0$, there exists a radial derivative, $v_r(x)$, of $v(x)$ and

$\varepsilon^{-1}|v((1+\varepsilon)x) - v(x)| \leq q_o(x) \in Q_\beta(R^{(0)})$ for $0 < \varepsilon < \varepsilon_o$ and some $\beta > 0$;

V obeys the "property of repulsivity"

(3) $v_r(x) \leq 0$ for $x \in R^{(0)}$, $x \neq 0$.

Thus, if either (I) or (II) is satisfied for the potentials, then Condition A is obeyed.

It follows from the results of Section 6 that

$$|L^a(\lambda \pm i0)|_{B^a \to B^a} \to 0 \qquad \text{as} \qquad |v|_o^{'}s \to 0$$

Thus, if V's are so small (in the corresponding norm) that $|L^a(\lambda \pm i0)| < 1$ then Condition B is also satisfied.

In Case (II) the restriction of Condition B is obeyed for all $\underline{a}$ with $k(a) \leqslant n-2$, since $\bar{A} \cap A_{k(a)+1} = \emptyset$ if $k(a) \leqslant n-2$. If this restriction is not obeyed for some $u \in A_{n-1}$ and some potential V_α, $u_\alpha = u$, then it is satisfied if V_α is replaced by $(1+\varepsilon)V_\alpha$ with any small nonzero ε [3].

3.3. Aside from these two cases, only limited results are obtained about a point spectrum imbedded into a continuous one. If V satisfy (1), (2) and $rv_r(x) \leqslant -\gamma v(x)$ $(0 < \gamma < 2,\ x \neq 0)$, then H has no eigenvalue in $[0,\infty)$ [20].

If H obeys a group of symmetry then its spectrum is composed of the spectrum of its parts on the invariant subspaces corresponding to the different types of irreducible representations of this group. Then, as was shown in [22, 23], in some cases isolated eigenvalues of one type of symmetry are imbedded into the continuous spectrum of another type. This unpleasant situation can be avoided if consideration is restricted to the invariant subspaces*. If the systems contain identical particles, then H obeys the corresponding symmetry group and, in order to carry out our considerations in the subspaces of definite types of irreducible representations of the group of permutation of the identical particles, one has to replace Berezin's equations,

---

* There is also another way to avoid this situation. As was shown in [24-26], eigenvalues imbedded into an essential spectrum as a rule are not stable under suitable small perturbation. Thus, one can conjecture that under proper small perturbation which breaks the symmetry the isolated eigenvalues of one type of symmetry imbedded into the continuous spectrum of another type disappear.

which are not invariant under permutation of the identical particles, by Fadeev-Yakubovsky equations which are invariant. The last equations have a structure similar to those of Berezin's and methods developed here can be applied to them.

Nothing is known about the existence of negative eigenvalues of some type of symmetry on the continuous spectrum of the same type of symmetry when V's do not satisfy to (I) and (II). However, as was noted in the first part of the footnote on the previous page, they are supposed to be unstable under small perturbations.

But, indirect condition A can be avoided by imposing a new direct restriction on the potentials. Namely, if the potentials V's are dilation analytic [27, detailed discussion see in 28], i.e., the operators

$$V_{ij}(\theta)f = V_{ij}(e^{\theta}x)f(x), \quad \theta \in R, \quad x \in R^3,$$

have Δ-compact continuation in a strip, $\{z \in C, \ |\mathrm{Im} z| < a\}$, one can separate, for any $a \in A$, the essential spectrum of H^a and its point spectrum embedded into the essential one, and then apply the technique developed here.

3.4 Condition B appears to be more essential than Condition A. As was shown in [29], if the condition of type B fails for a three body system then the discrete spectrum of such a system may be infinite, even for smooth potentials with compact supports (Effimov's effect). The points in the space of the potentials, V, where Condition B fails, have the character of "branch points": in these points the system changes the number of channels (a new eigenvalue for a subsystem appears). Thus, when Condition B fails the system may have qualitatively different character.

It is proved in [19] that

For any $V = \sum V_{ij}$ satisfying (2.11) and (2.12) a positive number ε_o exists such that $H(\varepsilon) = H + \varepsilon \sum_{u_\alpha \leq a} V_\alpha$ has no quasibound states at the left extremal point $\lambda^{a,1}$, $\lambda^{a,1} = \min_{b\in A'\cap \bar{A}_\varepsilon} \lambda^{b,1}(\varepsilon) = \min_{b\in A'\cap\bar{A}} \lambda^{b,1}$, of the continuous spectrum of $H(\varepsilon)$ for any ε, $0 < |\varepsilon| \leq \varepsilon_o$. Here $\{\lambda^{b,m}(\varepsilon),\ m = 1,2,\ldots, b\in A'\cap \bar{A}_\varepsilon\}$ is the threshold set and $\bar{A}_\varepsilon$ is the $\bar{A}$-set for $H(\varepsilon)$. Taking into consideration the group of symmetry one can carry out this result for all restrictions of H in the invariant subspaces.

3.5. The picture becomes much more clear if the additional restriction of the dilation analiticity (in the configuration space) is assumed. This condition gives a particular advantage when one considers a point spectrum imbedded into a continuous one.

There are already a number of the results about the systems with dilation analitic potentials:

(i) If $(-\Delta+1)^{-1/2} v_{ij}(-\Delta+1)^{-1/2}$ are compact and $(-\Delta+1)^{-1/2} v_{ij}(\theta)\ (-\Delta+1)^{-1/2}$, $\theta \in R$, have extensions to the strip $\{\theta: |\mathrm{Im}\ \theta| \leq \frac{1}{2}\}$ which are regular (continuous and analitic on the interior), then H has no positive eigenvalues [30].

(ii) If $v_{ij}(x)$ fall off at infinity faster than $|x|^{-2}$ and satisfy the condition similar to Condition B, then the point spectrum of H is finite [21].

(iii) The point spectrum of H imbedded into the continuous one is unstable under small perturbations [I.M. Sigal, unpublished].

(iv) Quasibound states are not stable under small perturbations [19], namely, for any (a,m), $a \in \bar{A} \cap A_2$ there exists $\varepsilon_o > 0$ such that $\{(b,k),\ b \in \bar{A}\cap A_2,\ H(\varepsilon) = H + \varepsilon \sum_{u_\alpha a} V_\alpha$ has no quasibound states at $\lambda^{b,k}(\varepsilon)\} = \{b,k),\ b \in \bar{A} \cap A_2,\ H$ has no quasibound states at $\lambda^{b,k}\}\cup\{(a,m)\}$ $\forall\ \varepsilon, 0 < |\varepsilon| \leq \varepsilon_o$,

where $\{\lambda^{b,k}(\varepsilon),\ k = 1,2,\ldots, b \in \bar{A}\}$ is the threshold set for the operator

$H(\varepsilon)$.

3.5. THEOREM. Let Conditions A and B be satisfied. Then the operator $R(z)$ may be expressed as

$$R(z) = \sum_{a\varepsilon A} \{P_a(z) + \overline{R}^a(z - \tau(P_a)) \otimes E_a\}, \tag{3.1}$$

where $\overline{R}^a(z) = 0$ for $a \in A_n$, and the operators $\overline{R}^a(z)$, $z \in \Pi_1^a$, $a \in A \setminus A_n$, are defined on $H_{\nu,\eta}(R^a)$, $\nu > 0$, $\eta > \frac{3}{2}$ with the properties

$$\overline{R}^a(z)\phi = \sum_{\substack{b \subseteq a, k \\ b\varepsilon A}} \frac{\psi^{b,k}(p^b) g^a_{b,k}(p^a_b; z, \phi)}{\lambda^{b,k} + \tau(p^a_b) - z}, \tag{3.2}$$

$$|g^a_{b,k}(z, \phi)|_{H_{s,\theta}(R^a_b)} + |\Delta|^{-s+t} |g^a_{b,k}(z + \Delta, \phi) - g^a_{b,k}(z, \phi)|_{H_{t,\theta}(R^a_b)} \tag{3.3}$$

$$\leq C|\phi|_{H_{\nu,\eta}(R^a)}(1 + |z|)^\delta ,$$

where $b \subset a$, $\delta > 0$, $s < \min(\nu, \frac{1}{2})$, $t < s$, $\theta < \eta$.

It follows from (3.1), (3.2) that the kernel $R(p,p'; z)$ of $R(z)$ has the form:

$$R(p,p'; z) = \sum_{a\varepsilon \overline{A}} \sum_{m=1}^{m(a)} \frac{\psi^{a,m}(p^a)\overline{\psi}^{a,m}(p^{a'})}{\lambda^{a,m} + \tau(p_a) - z} \delta(p_a - p_a')$$

$$+ \sum_a \sum_{\substack{b,c \subseteq a \\ b,c\varepsilon \overline{A} \\ k,\ell}} \frac{\psi^{b,k}(p^b)}{\lambda^{b,k} + \tau(p_b) - z} \Pi^a_{b,c;k,\ell}(p^a_b, p^{a'}_c; z - \tau(p_a)) \times \tag{3.4}$$

$$\times \frac{\overline{\psi}^{c,\ell}(p^{c'})}{\lambda^{c,\ell} + \tau(p_c) - z} \delta(p_a - p_a').$$

There is a relation between the limit of the resolvent, $R(z)$, on the continuous spectrum and the generalized eigenfunctions with given asymptotic behavior. Scattering theory is interested in the eigenfunctions of the operator H, whose asymptotes have the form $\Phi_{a,m}(p; q_a) = \psi^{a,m}(p^a)\delta(p_a - q_a)$.

Consider the limits

$$\psi^{\pm}_{a,m}(p;\ q_a) = \lim_{\varepsilon\to 0} i\varepsilon \int R(p,p';\ \lambda^{a,m} + \tau(p_a') \mp i\varepsilon)\Phi_{a,m}(p',q_a)dp'$$

$$= \lim_{\varepsilon\to 0} i\varepsilon \int R(p;\ p^{a'},q_a;\ \lambda^{a,m} + \tau(q_a) \mp i\varepsilon)\psi^{a,m}(p^{a'})dp^{a'}. \tag{3.5}$$

The existence of these limits follows owing to properties of the resolvent stated in Th. 3.5. (the detail deduction of this statement as well as the statements listed below is given in Sec. 7). Using the equation

$$H\ R(z) = E + zR(z), \tag{3.6}$$

one readily shows that they are the generalized eigenfunctions of H belonging to the point $\lambda^{a,m} + \tau(q_a)$ of the spectrum. A few manipulations of (3.5) yield an explicit expression for them in terms of the resolvent kernel:

$$\psi^{\pm}_{a,m}(p;\ q_a) = \psi^{a,m}(p^a)\delta(p_a - q_a) +$$

$$+ \sum_{\substack{b\in\bar{A},\ b\subset c,\\ c\supset a}} \frac{\psi^{b,\ell}(p^b)}{\lambda^{b,\ell} + \tau(p_b^c) - \lambda^{a,m} - \tau(q_a^c) \pm i0} \times$$

$$\times\ \Pi^c_{b,a;\ell,m}(p_b^c,q_a^c;\ \lambda^{a,m} + \tau(q_a^c) \mp i0)\delta(p_c - q_c) \tag{3.7}$$

This expression and Theorem 3.5 show that $\int \psi^{\pm}_{a,m}(p;\ q_a)f(p)dp$ is a Holder-continuous function of q_a for any $H_\nu(R)$, $\nu > 0$, $1 \leq m \leq m(a)$, $a \in \bar{A}\setminus A_2$.

3.6. We proceed as is sometimes done in spectral theory (see, e.g., [3]), viewing the generalized eigenfunctions (3.7) as components of the kernel of a unitary operator between two spaces. Normalization, orthogonality and completeness of generalized eigenfunctions will then be equivalent to the unitarity of this operator.

We define a space

$$\widehat{\mathscr{A}} = \sum_{a\in\bar{A},m} \oplus L_2(R_a), \quad (\hat{f},\hat{f}') = \sum_{a\in\bar{A},m} (f_{a,m}, f'_{a,m}), \tag{3.8}$$

and an operator defined in $\widehat{\mathscr{A}}$:

$$\hat{H}\hat{f} = \{(\lambda^{a,m} + \tau(p_a))f_{a,m}(p_a),\ a \in \bar{A},\ m\}, \quad \hat{f} = \{f_{a,m}(p_a),\ a \in \bar{A}, m\}.$$

Consider the expressions

$$g^{\pm}_{a,m}(p_a;\ f) = \lim_{\varepsilon\to 0} -i\varepsilon \int \bar{\psi}^{a,m}(p^a)(R(\lambda^{a,m} + \tau(p_a) \pm i\varepsilon)f)(p)dp^a \tag{3.9}$$

THEOREM.*) The expressions (3.9) define unitary operators

$$W_{\pm}f = \{g^{\pm}_{a,m}(p_a,f),\ a \in \bar{A},\ m\}$$

from $L_2(R)$ to $\widehat{\mathscr{A}}$. These operators satisfy the equalities

$$W_{\pm}H = \hat{H}\, W_{\pm}\ . \tag{3.10}$$

Now define operators by

$$W^{(\pm)}_{a,m}f = g^{\pm}_{a,m}(f). \tag{3.11}$$

3.7. It now remains to relate the result obtained in the stationary version of the problem to the solution of the nonstationary problem formulated in the Introduction, in other words, to determine the relationship between the operators $W^{(\pm)}_{a,m}$ of (3.11) and the wave operators $U^{(\pm)}_{a,m}$ of Sec. 2.

THEOREM. The operators $U^{(\pm)}_{a,m}(t) = e^{iHt}e^{-iH_{a,m}t}$ have strong limits $U^{(\pm)}_{a,m}$ in the space $\mathscr{A}_{a,m}$ as $t \to \pm\infty$. The limit operators $U^{(\pm)}_{a,m}$ satisfy the equalities

---

*) A statement of this result in the second quantization formalism may be found in [31].

$$U^{(\pm)}_{a,m}\, \psi^{a,m} f = W^{(\pm)*}_{a,m} f, \quad f \in L_2(R_a). \tag{3.12}$$

The scattering operator in the CM system is defined by [12]

$$S = W_+ W_-^* .$$

Theorem 3.6 implies that S is a unitary operator in $\widehat{\mathcal{H}}$ and commutes with the operator $\hat{H}$.

3.8. An outline of the proof of Theorem 3.5 now follows. As was mentioned in Introduction, the main tool in the study of behaviour of the resolvent $R(z)$ near the real axis is a construction and an investigation of an appropriate equation,

$$R(z) + L(z)R(z) = F(z) , \tag{3.13}$$

for $R(z)$. We accept as a primary criterion for the choice of the candidates for such equations the following requirements:

(1) The operators $L(z)$ and $F(z)$ can be expressed through the operators simpler than $R(z)$, for example, $R_a(z)$, $a \in A'$, and V_α.

(2) $L(z)$ is a compact operator on $D(H_o)$ unless $z \in [\mu,\infty)$;

(3) $F(z)$ is a bounded operator from $L_2(R)$ into $D(H_o)$ and has a bounded inverse (from $D(H_o)$ into $L_2(R)$)*) for $z \in C \setminus [\mu,+\infty)$.

---

*) Note that, since

$$F(z)(H - zI) = I + L(z)$$

and $L(z)$ is compact, $F(z)$ is a regularizer for $H - zI$ (in the sense the theory of partial differential equations).

It follows from (3) that

$$F^{-1}(z)(I + L(z)) = H - zI \tag{3.14}$$

and therefore if $\tilde{R}$ is a bounded operator in $L_2(R)$ with range in $D(H_o)$, satisfying equation (3.13) for some complex z, then $\tilde{R} = R(z)$. Hence one can define the boundary values of $R(z)$ when z approaches the real axis by a solution of this equation for $z = \lambda \pm i0$, $\lambda \in R$, if the latter exists and satisfies some continuity condition in z. The desired estimates on $R(z)$ are obtained if one solves (3.13) in an appropriate space.

Then we find the scales, B and $\hat{B}$ of Banach spaces and the linear map $\pi(z)$ of $\hat{B}$ into a scale of linear spaces such that $\pi(z)$ is an operator of a simple form and the operators $L(z)$ and $F(z)$ can be represented as follows

$$L(z)\pi(z) = \pi(z)\hat{L}(z)$$

and

$$F(z) = \pi(z)\hat{F}(z) ,$$

where the operators $\hat{L}(z)$ and $\hat{F}(z)$ have the properties

(i) $\hat{L}(z)$ is bounded in $\hat{B}$ and converges strongly as z approaches the real axis,

(ii) $\hat{L}(z)$ is compact for all $z \in C\setminus[\mu,+\infty)$;

(iii) A power of $\hat{L}(z)$ has compact boundary values at the real axis;

(iv) $-1 \in \sigma(\hat{L}(z))$, $z \neq \lambda^{a,m} \pm i0$, $a \in \bar{A} \cap A_2$, implies $z \in \sigma_p(H)$. z, here, is supposed to assume the values on the cut plane $C \setminus [\mu,+\infty)$ plus the values on both sides of the cut $[\mu,+\infty)$;

(v) $\hat{F}(z)$ is bounded from B to $\hat{B}$ and converges strongly as z approaches the real axis.

Note that (i) and (v) show that $\pi(z)$ absorbs all essential singularities of the functions $L(z)f$ and $F(z)f$, which arise as z approaches $[\mu,+\infty)$.

It follows from (i) - (v), that the resolvent $R(z)$ can be represented as

$$R(z) = P(z) + \pi(z)\hat{R}(z) ,$$

where $P(z) = P^a(z)$ $(a \in A_1)$ and $\hat{R}(z)$ is a bounded operator from B into $\hat{B}$ for all $z \in C\setminus[\mu,+\infty)$ which converges strongly as z approaches $[\mu,+\infty)$ with exception may be of some points of the set $\{\lambda^{a,m}, m = 1,2,\ldots,a \in \bar{A} \cap A_2\}$.

REMARKS. 1. (iv) is intimately related to the property of invertability of $F(z)$. Property (iv) is stronger than that we need. What we need really is to find $\hat{R}(\lambda \pm i0)$ (i.e. to invert $\hat{I} + \hat{L}(\lambda \pm i0)$) for all $\lambda \in [\mu,+\infty)$ up to a countable set. However, in the proof of the statement about $R(z)$ it is necessary to exclude the exceptional sets in the induction hypothesis about the operators $R^a(z)$, $a \in A'$. The property (iv) written for the decompositions $a \in A'$ and Conditions A and B provide such an exclusion in the proof of the induction hypothesis.

2. The main consequences of (i) - (v) remain the same if $L(z)$ and $F(z)$ are unbounded operators and even if $L(z)$ has a nondense domain [32]. In our case the proof of (i) is a part of the proof of (iii).

(v) is proved in the same way as (i). The proof of (ii) is simple, since in this case z belongs to the resolvent sets of H_a, $a \in A'$.

Now we outline briefly the proof of (iii). $\hat{L}(z)$ can be represented as a finite linear combination of monomials of the form

$$\Pi\, [\hat{R}'_{a_i}(z) V_{\alpha_i} \pi(z)] \tag{3.15}$$

where $\pi(z)\hat{R}'_a(z) = R_a(z)$, $a \in A'$. We prove that $\hat{L}(z)$ is a bounded operator in $\hat{B}$, continuous in all factors, $\hat{R}'_a(z)$'s and V_α's, contained in its monomials, in the uniform operator topology. This is the most complicated part of the proof. It appears in the first half of Section 6 and in Appendix II. Then we approximate $\hat{R}'_a(z)$'s and V_α's in the monomials by operators with good kernels. The approximating operators, $\hat{L}_k(z)$, converge to $\hat{L}(z)$ in the uniform operator topology. Given a suitable approximation of $\hat{R}'_a(z)$'s and V_α's, the operators $L^3_k(z)$ are shown to converge in the operator norm as z approaches the real axis. Hence $\hat{L}^3(z)$ converges as $\operatorname{Im} z \to \pm 0$ in the uniform operator topology too and therefore has compact boundary values at R. In order to demonstrate the desired properties of $\hat{L}^3_k(z)$, $k = 1,2,\ldots$, we estimate its kernel. The latter is representable in the form

$$\int \frac{\phi(p,q,k)\,dk}{\prod_{i=1}^{s} [P_i(p,q,k)-z]}, \quad k \in R^{3m}, \quad m \geq 2(n-1), \tag{3.16}$$

where $P(\cdot)$'s are polynomials of the second degree. In order to estimate such an integral we use the well-known Feynman's identity

$$\prod_{i=1}^{s} A_i^{-1} = \int_0^1 \Big(\sum_{i=1}^{s} \alpha_i A_i\Big)^{-s} \delta\Big(\sum_{i=1}^{s} \alpha_i - 1\Big) \prod_{i=1}^{s} d\alpha_i ,$$

which reduces (3.16) to an integral with one polynomial for the second degree in the s-th power in the denominator. After this we integrate the obtained integral by parts. This gives us the desired estimates.*)

---

*) Integrals of the type (3.16) can be estimated also by the method proposed earlier in [4]. This method uses the "time representation" of the denominator

$$\prod_{i=1}^{s} A_j^{-1} = \int_0^\infty e^{i \sum_{j=1}^{s} A_j t_j} \prod_{j=1}^{s} dt_j , \qquad \operatorname{Im} A_j > 0 ,$$

instead of the Feynman's identity.

In the conclusion of this outline we note how the structure of the scale $\hat{B}$ and the map $\pi(z)$ can be guessed. We consider the operator $R_a(z)$ such that $\underline{a} \in \bar{\bar{A}} \setminus A_n$. By (2.22) and (2.22')

$$R_a(z)f = R^a(z - \tau(p_a))f(p_a) \tag{3.17}$$

The structure of $R^a(z)$ near the discrete spectrum of H^a is given by the well-known theorem of Functional Analysis:

$$R^a(z) = \sum(\lambda^{a,m} - z)^{-1} P^{a,m} + R_1^a(z) ,$$

where $R_1^a(z)$ is an analitic operator outside of the essential spectrum of H^a. Hence by (3.17) we obtained for $R_a(z)$ the following representation

$$R_a(z) = P_a(z) + R_{1,a}(z) ,$$

$$R_{1,a}(z)f = R_1^a(z - \tau(p_a))f(p_a) .$$

Thus since $L(z)$ is represented as a finite linear combination of monomials of the type (3.15) the vector $L(z)f$ for any $f \in D(H_o)$ has the form

$$L(z)f = \sum_{a\varepsilon \quad ,m} \psi^{a,m} R_{a,m}(z)\, g_{a,m}(z,f) + \text{something},$$

where $\psi^{a,m}$ is the operator of multiplication on $\psi^{a,m}(p^a)$, $R_{a,m}(z) = (\lambda^{a,m} + T_a - z)^{-1}$ and $g_{a,m}(z,f) \in L_2(R_a)$, some functions. This expression suggests $\hat{B}$ and $\pi(z)$ of the form

$$\hat{B} = \bigoplus_{a\varepsilon\bar{A}\ ,m} B_{a,m} , \qquad B_{a,m} \subset L_2(R_a)$$

and

$$\pi(z)\hat{f} = \sum \psi^{a,m} R_{a,m}(z) f_{a,m} , \qquad \hat{f} = \{f_{a,m}\} \in \hat{B} .$$

4. EQUATION FOR THE RESOLVENT OF H

In this section we derive the equations with whose help we intend to study the resolvent $R(z)$ of H. We set

$$R_o(z) = (H_o - zE)^{-1}.$$

The resolvent satisfies the equation

$$R(z) + R_o(z)VR(z) = R_o(z). \tag{4.1}$$

Equation (4.1) has the desired properties (I), (II) (see sec. 3) only in the case $n = 2$. To deal with the case $n > 2$, we must reorganize the equation in some way; this we now proceed to do.

Equation (4.1) may be written

$$R(z) + \sum_{i<j} R_o(z)V_{ij}R(z) = R_o(z). \tag{4.2}$$

Define operators by

$$L_a(z) = R_o(z)V_{ij}, \quad a = u_{ij}, \tag{4.3}$$

In terms of these operators Eq. (4.2) may be written as

$$R(z) + \sum_{a_1 \in A_{n-1}} L_{a_1}(z)R(z) = R_o(z). \tag{4.4}$$

If $f + L_a(z)f = g$, $f \in D(H_o)$, $a \in A_{n-1}$, $\operatorname{Im} z \neq 0$, then $g \in D(H_o)$ and $(H_a - zE)f = (H_o - zE)g$. Hence it follows that the operators $E + L_a(z)$, $a \in A_{n-1}$, $\operatorname{Im} z \neq 0$, are invertible. Inverting the operator $E + L_{a_1}(z)$ in (4.4) and introducing the notation

$$B_{a_1}(z) = (E + L_{a_1}(z))^{-1} - E,$$

we obtain

$$R(z) + [L_{a_2}(z) + L_{a_3}(z) + \dots + B_{a_1}(z)(L_{a_2}(z) + L_{a_3}(z) + \dots)]R(z)$$

$$= (E + L_{a_1}(z))^{-1} R_o(z).$$

We now invert $E + L_{a_2}(z)$, iterating the procedure until all the operators $E + L_{a_i}(z)$ have been inverted. The final result is

$$\prod_{a_i \varepsilon A_{n-1}} (E + L_{a_i}(z))^{-1} (E + \sum_{a_i \varepsilon A_{n-1}} L_{a_i}(z))R(z) = \prod_{a_i \varepsilon A_{n-1}} (E + L_{a_i}(z))^{-1} R_o(z). \tag{4.5}$$

Introducing the notation

$$B_a(z) = (E + L_a(z))^{-1} - E, \quad a \in A_{n-1}, \tag{4.6}$$

we rewrite Eq. (4.5) as follows:

$$R(z) + \sum_{\substack{a_j \varepsilon A_{n-1}, s \geq 1 \\ i_1 > \dots > i_s, i > i_s}} B_{a_{i_1}}(z) \dots B_{a_{i_s}}(z) L_{a_i}(z) R(z)$$

$$= \prod_{a \varepsilon A_{n-1}} (E + L_a(z))^{-1} R_o(z). \tag{4.7}$$

This form of the equation provides for convenient grouping of the terms in coefficient of $R(z)$. Equation (4.7) satisfies conditions (I), (II) of sec. 3 if $n = 3$. If $n > 3$, however, it has the same shortcomings as Eq. (4.1) for $n > 2$. We must therefore proceed further with our reorganization of Eq. (4.1).

In the second term on the left of Eq. (4.7), combine all summands corresponding to the same partitions:

$$R(z) + \sum_{\substack{a\varepsilon \bigcup_1^{n-2} A_k}} \sum_{\substack{a_i \varepsilon\, n-1,\, s\geq 1 \\ i_1>\ldots>i_s,\, i>i_s \\ \cup a_{i_k} \cup a_i = a}} B_{a_{i_1}}(z)\ldots B_{a_{i_s}}(z) L_{a_i}(z) R(z)$$

$$= \prod_{a\varepsilon\, A_{n-1}} (E + L_a(z))^{-1} R_o(z).$$

Now denote the sums of all summands corresponding to partitions $a \in A_{n-2}$ by $L_a(z)$, and the sum of the remaining summands by $C(z)$. Then Eq. (4.7) becomes

$$R(z) + [\sum_{a\varepsilon A_{n-1}} L_a(z) + C(z)]R(z) = \prod_{a\varepsilon A_{n-1}} (E + L_a(z))^{-1} R_o(z). \tag{4.8}$$

We now invert the operators $E + L_a(z)$, $a \in A_{n-2}$, in turn, as done previously for the operators $E + L_a(z)$, $a \in A_{n-1}$, and so on. The (n - 1)-th iteration of this procedure yields an equation satisfying conditions (I), (II) of Sec. 3:

$$R(z) + L(z)R(z) = F(z), \tag{4.9}$$

where the operator $L(z)$ is defined inductively:

$$L_a(z) = R_o(z)V_{ij}, \quad a = u_{ij} \in A_{n-1}, \tag{4.10}$$

$$L_a(z) = \prod_{\substack{b\subset a \\ b\varepsilon A_{k(a)+1}}} (E + L_b(z))^{-1} \ldots \prod_{\substack{v\subset a \\ v\varepsilon A_{n-1}}} (E + L_v(z))^{-1}$$

$$(E + \sum_{\substack{v\subset a \\ v\varepsilon A_{n-1}}} L_v(z)) - E,$$

$$L(z) = L_a(z) \quad (a \in A_1),$$

and the operator $F(z)$ is defined by

$$F(z) = \prod_{a\varepsilon A_2} (E + L_a(z))^{-1} \cdots \prod_{v\varepsilon A_{n-1}} (E + L_v(z))^{-1} R_o(z). \qquad (4.11)$$

The whole of the above procedure could have been applied, not to the equation (4.1) for the resolvent of the total Hamiltonian H, but instead to the equation for the resolvent $R_a(z)$ of the Hamiltonian H_a of a compound system. The resulting equation (instead of (4.9)) would then have been

$$R_a(z) + L_a(z) R_a(z) = F_a(z), \qquad (4.12)$$

where

$$F_a(z) = \prod_{\substack{b\subset a \\ b\varepsilon A_{k(a)+1}}} (E + L_b(z))^{-1} \cdots \prod_{\substack{v\subset a \\ v\varepsilon A_{n-1}}} (E + L_v(z))^{-1} R_o(z). \qquad (4.13)$$

This equation will enable us to derive the necessary information about the behavior of the operator $R_a(z)$ as z approaches the real axis. Note that the operators $L_a(z)$ depend on the order in which the operators $E + L_b(z)$, $b \subset a$, $b \in A_s$, are inverted. To fix our ideas, we stipulate that the products of operators $E + L_b(z)$, $b \subset a$, $b \in A_s$, $k(a) < s \lesssim n - 1$, are evaluated in the order specified at the end of Sec. 2.

We now establish a formula for the operators $L_a(z)$ which will be useful in the sequel.

We first set

$$B_a(z) = (E + L_a(z))^{-1} - E \quad \text{for} \quad a \in A . \qquad (4.14)$$

Using these operators, one proves by induction (omitting the argument z in the following formulas) that

$$L_a = \sum B_{b_{i_1}} \cdots B_{b_{i_s}} \cdots B_{c_{j_1}} \cdots B_{c_{j_t}} B_{d_{\ell_1}} \cdots B_{d_{\ell_{p-1}}} L_{d_{\ell_p}} , \qquad (4.15)$$

where the summation extends over all partitions such that

$$i_1 > \dots > i_s, \dots , j_1 > \dots > j_t, \ell_1 > \dots > \ell_{p-1}, \ell_p > \ell_{p-1},$$

$$k(a)+1 \leqslant k(b_{i_1}) = \dots = k(b_{i_s}) < \dots < k(c_{j_1}) = \dots = k(c_{j_t})$$
$$< k(d_{\ell_1}) = \dots = k(d_{\ell_p}), \tag{4.16a}$$

$$b_{i_j} \cup \dots \cup d_{\ell_p} = a .$$

The definition (4.10) now implies the formula (z is still omitted)

$$(E + L_a)^{-1} = (E + \sum_{\substack{v\varepsilon A_{n-1} \\ v\subset a}} L_v)^{-1} \prod_{\substack{v\varepsilon A_{n-1} \\ v\subset a}}{}' (E + L_v) \dots \prod_{\substack{b\varepsilon A_{k(a)+1} \\ b\subset a}}{}' (E + L_b)$$

$$= E + (E + \sum_{\substack{v\varepsilon A_{n-1} \\ v\subset a}} L_v)^{-1} \sum_{\substack{k(f\cup\dots\cup g)\leqslant n-2 \\ k(f)\geqslant \quad \geqslant k(g) \\ f\cup\dots\cup g\subseteq a}}{}' L_f \dots L_g . \tag{4.17}$$

On the other hand, we have

$$(E + L_a)^{-1} = E - (E + L_a)^{-1} L_a . \tag{4.18}$$

Comparing these two formulas we get

$$(E + L_a)^{-1} = E + (E + \sum_{\substack{v\varepsilon A_{n-1} \\ v\subset a}} L_v)^{-1} \sum_{\substack{f\cup\dots\cup g=a \\ k(f)\geqslant\dots\geqslant k(g)}}{}' L_f \dots L_g, \tag{4.19}$$

where the prime both here and in Eqs. (4.21), (4.22) below means that the summation is performed only over terms occurring in the expression

$$\prod_{\substack{v\in A_{n-1}\\ v\subset a}}{}' (E + L_v) \cdots \prod_{\substack{b\in A_{k(a)+1}\\ b\subset a}}{}' (E + L_b). \tag{4.20}$$

We now convert the second term in (4.19) to a form more convenient for investigation. Comparing (4.19) and (4.17), we see that for any b

$$-\sum_{\substack{\phi\cup\ldots\cup\psi\subseteq b\\ k(\phi)\geq\ldots\geq k(\psi)}}{}' L_\phi \cdots L_\psi L_b = \sum_{\substack{f\cup\ldots\cup g=b\\ k(f)\geq\ldots\geq k(g)}}{}' L_f \cdots L_g. \tag{4.21}$$

Applying these identities successively to the sum in the second term of (4.19), we obtain

$$B_a(z) = R_a(z)(H_o - zE)\sum{}' L_{f_1}(z) \cdots L_{f_t}(z), \tag{4.22}$$

where the summation extends over all partitions such that $f_1 \in A_{n-1}$, $k(f_s) \geq k(f_{s+1})$, $k(f_t) \geq k(a) + 1$, $f_i \not\subseteq f_{s+1}$, $s = 1,\ldots,t-1$, $\cup f_i = a$ (and the convention for primed summation symbols adopted above is also observed).

Applying to (4.13) the formula

$$(E + L_b)^{-1} \prod_{s=k(b)+1}^{n-1} \prod_{\substack{c\subset b\\ c\in A_s}} (E + L_c)^{-1} R_o(z) = (E + L_b)^{-1} F_b(z) = R_b(z),$$

we obtain the following convenient formulas:

$$F_a(z) = \sum_{b\ a} \tilde{R}_b(z) + M_a(z), \tag{4.23}$$

$$\tilde{R}_b(z) = \begin{cases} R_o(z), & k(b) = n \\ R_b(z) - \sum\limits_{c\subset b} \tilde{R}_c(z), & k(b) < n \end{cases} \tag{4.24}$$

$$M_a(z) = \sum B_{f_t}(z) \ \ldots \ B_{f_1}(z)R_o(z) \tag{4.25}$$

where the range of the summation symbol is $f_1 \in A_{n-1}$, $k(f_s) \geq k(f_{s+1})$, $k(f_t) \geq k(a) + 1$, $f_i \not\subseteq f_{s+1}$, $s = 1,\ldots,t-1$, $\cup f_i = a$.

To end this section, we observe that the operators $R_{b,a}(z)$, $b \subseteq a$, introduced in Sec. 2 satisfy the equations

$$R_{b,a}(z) + L_{b,a}(z)R_{b,a}(z) = F_{b,a}(z), \tag{4.26b}$$

where

$$L_{b,a}(z) = \prod_{s=k(b)+1}^{n-1} \prod_{\substack{c \subset b \\ c\varepsilon A_s}} (E^a + L_{c,a}(z))^{-1}(E^a + \sum_{\substack{v \subset b \\ v\varepsilon A_{n-1}}} L_{v,a}(z)) - E^a, \tag{4.27b}$$

$L_{v,a}(z) = R_{o,a}(z)\, V^a_{ij}$, $v = \{(ij) \prod_{k\neq i,j} (k)\} \in A_{n-1}$,

$$F_{b,a}(z) = \prod_{s=k(b)+1}^{n-1} \prod_{\substack{c \subset b \\ c\varepsilon A_s}} (E^a + L_{c,a}(z))^{-1} R_{o,a}(z). \tag{4.28b}$$

By analogy with (4.15), (4.14), (4.22) - (4.25) we have formulas for $L_{b,a}(z)$ and $F_{b,a}(z)$:

$$B_{b,a} = \sum B_{b_{i_1},a} \ldots B_{b_{i_s},a} \ldots b_{c_{j_1},a} \ldots B_{c_{j_t},a} B_{d_{\ell_1},a} \ldots B_{d_{\ell_{p-1}},a} L_{d_{\ell_p},a}. \tag{4.29b}$$

where the first and second summations extend over the partitions satisfying conditions (4.16b),

$$B_{b,a}(z) = (E + L_{b,a}(z))^{-1} - E^a$$

$$= R_{b,a}(z)(H^a_o - zE^a) \sum L_{f_1,a}(z) \ldots L_{f_t,a}(z) \ , \tag{4.30b}$$

where the summation extends over $f_1 \in A_{n-1}$, $k(f_s) \geq k(f_{s+1})$, $f_i \not\subseteq f_{s+1}$, $s = 1,\ldots,t-1$, $\cup f_i = b$, and

$$F_{b,a}(z) = \sum_{c\ B} \tilde{R}_{c,a}(z) + M_{b,a}(z), \tag{4.31b}$$

$$M_{b,a}(z) = \sum B_{f_t,a}(z) \ldots B_{f_1,a}(z) R_{o,a}(z), \tag{4.32b}$$

where the scope of the summation symbol is $f_1 \in A_{n-1}$, $k(f_s) \geq k(f_{s+1})$, $\cup f_i = b$, $f_i \not\subseteq f_{s+1}$, $s = 1,\ldots,t-1,$, $k(f_t) \geq k(b) + 1$,

$$\tilde{R}_{b,a}(z) = R_{b,a}(z) - \sum_{c<b} \tilde{R}_{c,a}(z) \tag{4.33b}$$

From (4.31) we obtain the following equations for $\tilde{R}_{b,a}(z)$, $b \subseteq a$:

$$\tilde{R}_{b,a}(z) + L_{b,a}(z)\, \tilde{R}_{b,a}(z) = \tilde{F}_{b,a}(z), \tag{4.34b}$$

where

$$\tilde{F}_{b,a}(z) = - L_{b,a}(z) \sum_{c \subset b} \tilde{R}_{c,a}(z) + M_{b,a}(z). \tag{4.35b}$$

Taking the adjoints in $L_2(R^a)$ of (4.31), (4.42), we obtain equations for $R_{b,a}(z)$ and $\tilde{R}_{b,a}(z)$:

$$R_{b,a}(z) + R_{b,a}(z)\, L'_{b,a}(z) = F'_{b,a}(z), \tag{4.36b}$$

$$\tilde{R}_{b,a}(z) + \tilde{R}_{b,a}(z)\, L'_{b,a}(z) = \tilde{F}'_{b,a}(z), \tag{4.37b}$$

where the operators $L'_{b,a}(z)$, $F'_{b,a}(z)$ and $\tilde{F}'_{b,a}(z)$ are the adjoints in $L_2(R^a)$ of $L_{b,a}(\bar{z})$, $F_{b,a}(\bar{z})$ and $\tilde{F}_{b,a}(\bar{z})$, respectively. Equation may also be derived directly, by reorganization of the equation

$$F_{b,a}(z) + R_{b,a}(z) \sum_{u_\alpha \subset b} V_\alpha^a R_{o,a}(z) = P_{o,a}(z),$$

using a procedure similar to that used for Eq. (4.9).

In the special case $b = a$, we shall use the notation

$$L^a(z) = L_{a,a}(z), \quad \tilde{R}^a(z) = R_{a,a}(z),$$

$$\tilde{F}^a(z) = \tilde{F}_{a,a}(z). \tag{4.38a}$$

5. RESOLVENT OUTSIDE THE CONTINUOUS SPECTRUM. EIGENFUNCTIONS OF THE DISCRETE SPECTRUM

5.1. This section is devoted to certain results, needed later (Sec. 6), concerning the behaviour of the resolvent $R^a(z)$ outside the continuous spectrum of the operator H^a. We shall also establish estimates on the eigenfunctions of the discrete spectrum of H^a, $a \in \bar{A}$. Similar results were first published as far as we know in [33].

Let $d_a(z)$ denote the distance from z to the interval $[\mu^a +\infty)$

$$[d_a(z)]^2 = [\operatorname{Im} z]^2 + [\min(0,\mu^a - \operatorname{Re} z)]^2, \quad \rho^a_\varepsilon = \{z, d_a(z) \geq \varepsilon\}$$

PROPOSITION. For any partition $\underline{a}$, the operator $\tilde{R}^a(z)$, defined by (4.26a), (4.29a), admits the folowing representation for $z \notin [\mu^a +\infty)$:

$$\tilde{R}^a(z) = P^a(z) + \overline{R}^a(z), \tag{5.1}$$

where the operator $\overline{R}^a(z)$ satisfies the estimates*)

$$|D^k_z \overline{R}^a(z)|_{H_{\nu,\eta}(R^a) \to R_{o,a}(i) H_{\nu_o,\eta_o}(R^a)} \leq C(d_a(z)+1)^{-k}, \; z \in \rho^a_\varepsilon, \quad k = 0,1,\ldots . \tag{5.2}$$

5.2. The proof is by induction on partitions $\underline{a}$. For partition in A_n the assertion is trivial: $\overline{R}^a(z) = 0$. Assuming that it is true for all $b, b \subset a$, we shall show that it is also true for $\underline{a}$.

To this end, we use Eq. (4.41a):

---

*) For the definition of $H_{\nu,\eta}(R^a_b)$ see p. 18.

$$\tilde{R}^a(z) + L^a(z)\, \tilde{R}^a(z) = \tilde{F}^a(z). \tag{5.3}$$

We note that the complement of the interval $[\mu^a\ +\infty)$ is a subset of the resolvent set for the operators H^b, $b \subset a$.

LEMMA. If z lies outside the continuous spectrum of H^a, then the operator $D^k L^a(z)$ is continuous from $R_{o,a}(-1) H_{\nu,\eta}(R^a), \nu \geq o, \eta > \frac{3}{2}$, to $R_{o,a}(-1)\ H_{\nu_o,\eta_o}(R^a)$ and its norm is bounded by $C(d_a(z)+1)^{-k}$, $z \in \rho^a_\varepsilon$, $\forall\ k$.

PROOF. From (4.29a), (4.30a) we see, that the operator $L^a(z)$ can be represented as a finite sum of operators

$$\prod_{i=1}^{k} [R_{b_i,a}(z) V^a_{\alpha_i}] \tag{5.4}$$

with b's and α's satisfying the conditions

$$\bigcup_{t=1}^{k} u_{\alpha_t} = a, b_i \subset \bigcup_{t=1}^{i} u_{\alpha_t} \tag{5.5}$$

We assume that the operators $R_{b,d}(z)$, $b \subset c \subset a$, satisfy the estimates

$$|R^{-1}_{o,d}(i) D^k_z R_{b,d}(z)\ f|_{H_{\nu_o,\eta_o}(R^d)} \leq C(d_b(z)+1)^{-k}\ .$$

$$|f|_{H_{\nu_o,\eta_o}(R^d)}\ , \qquad k = 0,1,\dots\ . \tag{5.6b,d}$$

for $|z - \sigma(H_b)|>\varepsilon$. In App. I the following estimates for $V^a_\alpha R_{o,a}(z)$, $z \in C/[0,\infty)$,

$$|D^k_z\ V^a_\alpha R_{o,a}(z) f|_{H_{\nu_o,\eta_o}(R^e) \times H_{\nu,\eta}(R^a_e)} \tag{5.7}$$

$$\leq Cd_o(z)^{-1-k}(1 + d_o(z))|f|_{H_{\nu_o,\eta_o}(R^d) \times H_{\nu,\eta}(R^a_d)}, k = 0, 1,\dots, d = e\ \ u_\alpha,\ e \subset a,$$

are proved. Here $d_o(z)$ is the distance from the point z to the semi-axis $(0,\infty)$. Applying the estimates (5.6 b,d), (5.7) to (5.4) we obtain immediately the desired estimates for $L^a(z)$.

5.3. According to the known theorems in functional analysis the resolvent $R^a(z)$ is represented as

$$R^a(z) = P^a(z) - R_1^a(z), \tag{5.8}$$

where $R_1^a(z)$ is an operator from $L_2(R^a)$ to $D(H^a)$, analytic outside the continuous spectrum of H^a. Hence,

$$\overline{R}^a(z) = \tilde{R}^a(z) - P^a(z) = R_1^a(z) - \sum_{b\subset a} \tilde{R}_{b,a}(z) \tag{5.9}$$

is also an operator from $L_2(R^a)$ to $D(H^a)$, analytic outside the continuous spectrum of H^a. (5.3), (5.9) imply for it the equation.

$$\overline{R}^a(z) + L^a(z)\,\overline{R}^a(z) = \tilde{F}_1(z) , \tag{5.10}$$

where

$$\tilde{F}_1^a(z) = \tilde{F}^a(z) - (E^a + L^a(z))\, P^a(z) \tag{5.11}$$

Let P^a be the projection operator onto the eigensubspace of the discrete spectrum of H^a. Using the equation

$$E^a + L^a(z) = F^a(z)\,(H^a - zE^a) \tag{5.12}$$

we transform the oprator (5.11) to the form

$$\tilde{F}_1^a(z) = \tilde{F}^a(z) - F^a(z)\, P^a$$

5.4. From Eq. (5.12) and lemma 5.2 follows:

LEMMA. Let ψ^a be an eigenfunction of the operator H^a with an eigenvalue λ, $\lambda < \mu^a$. Then ψ^a satisfies the equation

$$\psi + L^a(\lambda)\ \psi = 0 \tag{5.13}$$

and belong to the space $R_{o,a}(i)\ H_{\nu_o,\eta_o}(R^a)$

5.5. Using this lemma, we obtain the following statement in the same way as lemma 5.2 was obtained.

LEMMA. The operator $\tilde{F}_1^a(z)$ for $z \notin \sigma_e(H^a)$ satisfies the estimates (5.2a).

Lemmas 5.2, 5.5 and Eq. (5.10) imply the estimates (5.2a) for $\tilde{R}_1^a(z)$, $z \notin [\mu^a, +\infty)$.

5.6. To finish the proof it remains to show that the assumption on the operators $R_{b,d}(z)$, $b \subset d \subseteq a$, (see, 5.6 b,d) is valid. From (5.2) it follows (one can use unequality $N^c < N_a^c$, $a \subset c$, where $N^c = N_b^c$ $(b \in A_n)$, for this), that the operators $R_{a,c}(z)$, $a \subset c$, satisfy the estimates (5.6 a,c). For $a \in A_{n-1}$ the proof of this fact is complete, because $b \subset a \in A_{n-1} \to b \in A_n$, $R_{b,a}(a) = R_{o,a}(z)$ and we do not need the assumption about $R_{b,a}(z)$, $b \subset a$. Thus the estimates (5.6 b,a) for the operators $R_{b,a}(z)$ are proved by induction. Proposition 5.1 is proved.

5.7. We see from (4.28a) that for $z \notin [\mu^a, +\infty)$ the operator $F^a(z)$ has the bounded inverse from $D(H^a)$ into $L_2(R^a)$. Then from (5.12) follows the statement which is inverse to lemma 5.4:

LEMMA. (i) The homogeneous equation (5.1) has no solutions in $D(H^a)$ for nonreal z.

(ii) If Eq. (5.12) has a nontrivial solution $\psi(\lambda) \in D(H^a)$ for $z = \lambda$, $\lambda < \mu^a$, then λ is a point of the discrete spectrum of H^a, and $\psi(\lambda)$ is a corresponding eigenfunction.

In the conclusion of this section we write out for reference an estimate on the eigenfunctions of the discrete spectrum of H^a which follows from lemma 5.4:

$$|\psi^a|_a \equiv |R_{o,a}^{-1}(-1)\psi^a|_{H_{\nu_o,\eta_o}(R^a)} < C \tag{5.14}$$

6. RESOLVENT ON THE CONTINUOUS SPECTRUM

In this section we study the behaviour of the resolvent, $R(z)$, when the parameter, z, approaches the continuous spectrum, $\sigma_c(H)$. The main results of this section are contained in Proposition 6.3 which implies Theorem 3.5 of Section 3.

This section is organized as follows: First of all, we define the scales of Banach spaces and describe their properties, §§ 1,2. Then in §3 we formulate the main result of the section (Proposition 6.3). The proof of this result occupies the remaining paragraphs (except the last one). As was noted at the end of Section 3, the proof is conducted by induction on the decompositions $a \in A$. Because of this, the proposition is formulated in a form which permits one to state it as an induction hypothesis. The main tool of the proof is the study of Eqs. (4.26a) for the resolvents $R^a(z)$, $a \in A$, which are reduced to Fredholm equations of type

$$\hat{R}^a(z) + \hat{L}^a(z)\hat{R}^a(z) = \hat{F}^a(z)$$

(for the definitions and discussions see the end of Section 3). The task of §§ 5-23 is to prove the uniform boundedness and the strong continuity of the operator $\hat{L}^a(z)$, $z \in \Pi^a$, the most lengthy estimates are carried out in Appendix II).

In §§ 24-31 we show that $[\hat{L}^a(z)]^3$, $z \in \Pi^a$, is a compact operator in the chosen scale of Banach spaces. In §§ 25 - 28 we approximate $\hat{L}^a(z)$ by the operators $\hat{L}^a_n(z)$. In §§ 29,30 (all estimates are carried out in Appendix III) we prove that $[\hat{L}^a_n(z)]^3$ are compact operators for all $z \in \Pi^a$. The proof of the uniformity of the approximation is contained in

§ 31. This shows that the operator $[\hat{L}^a(z)]^3$ is compact $\forall z \in \Pi^a$ also. In § 32 we state the properties of the inhomogeneous term, $\hat{F}^a(z)$, in the equation for $\hat{R}^a(z)$. In §§ 33-36 we study the homogeneous equation

$$\hat{f} + \hat{L}^a(z)\hat{f} = 0$$

and show that if $\hat{f}_\lambda$ is a solution for $z = \lambda + i0$ (or $\lambda - i0$), where $\lambda \neq \lambda^{b,m}$, $m = 1,\dots,m(b)$, $b \in \bar{A} \cap A_{k(a)+1}$, in the chosen scale of Banach spaces then $\pi^a(\lambda + i0)\hat{f}_\lambda = \pi^a(\lambda - i0)\hat{f}_\lambda$ is an eigenfunction of H^a with eigenvalues λ (which is imbedded into the continuous spectrum of H^a when $\lambda \in \sigma_c(H)$). In § 37 we formulate the restriction on the operator H (Condition B) and state the properties of $(\hat{E}^a + \hat{L}^a(z))^{-1}$ under this restriction and others made before. This leads to the desired conclusions about the solutions of the equation under consideration. In § 38 the proof of the Lemma 6.15 is carried out. This completes the proof of Proposition 6.3.

To conclude this outline, we remark that we have chosen for the scales of the Banach spaces mentioned above the symmetric sums of the anisotropic, weighted $\Lambda_\sigma^{m,m}$-spaces (in the terminology of Stein [34]), which vary essentially through the indices σ and η, where σ shows the degree of uniform smoothness and η, the fall off at infinity (as a rule, we drop η in the notation for these spaces and introduce in the notation other index important for the sequential estimates).
The main reason for the choice of such spaces is to make the operators $\hat{L}^a(z)$ bounded (uniformly in $z \in \Pi^a$) and strongly continuous in $z \in \Pi^a$. The estimates which use the particular structure of these spaces in an essential way are contained in Appendix II.

Recall that we are omitting the index m in the eigenfunctions and eigenvalues of the operators H^a, $a \in \bar{A}$, and operating as if each H^a had only one eigenvalue.

6.1. Now we proceed to the description of Banach spaces which are used for the study of the equations for the resolvent derived in Section 4. With any pair (b,α), where b is a decomposition and α is a pair of indices from the set $\{1,\ldots,k(b)\}$, we associate the decomposition $b_\alpha \in A_{k(b)-1}$ defined as follows

$$b_\alpha = \{ \bigcup_{i\in\alpha} C_i \ , C_k \ , k \notin \alpha \ ; C_k \in b \}$$

The following operators are familiar from the theory of functional spaces and partial differential equations

$$J^\delta_{b,\alpha} f = \int e^{ip_b x_b} \hat{f}(x_b) \ (1+|x_b^{b_\alpha}|^2)^{-\delta/2} dx_b$$

$$\hat{f}(x_b) = \frac{1}{2\pi} \int e^{-ix_b p_b} f(p_b) \ dp_b$$

Let

$$\chi \in C^\infty \ , \quad \chi(s) = \begin{cases} 1 \ , & s > -\frac{1}{10n}\kappa \\ 0 \ , & s < -\frac{1}{5n}\kappa \end{cases} \ , \quad \kappa = \min_{b\ a} (\lambda^b - \lambda^a) \ ,$$

and

$$\chi_b(\gamma) = \chi(T^a_b - \gamma) \quad \text{if} \quad b \in \bar{A} \setminus A_n \quad \text{and} \quad = E^a_b \quad \text{if} \quad b \in A_n$$

We set (6.1)

$$H_\gamma(R^a_b) = H^m_{\nu,\gamma,\eta}(R^a_b) = \{ f \in N^a_b(\eta) \ L_m(R^a_b) \ : \ \chi_b(\gamma) f \in \Sigma \ J^\nu_{b,\alpha} N^a_b(\eta) \ L_m(R^a_b) \}$$

The norm in $H_\gamma(R_b)$ is defined as

$$|f|_{H_\gamma(R^a_b)} = |M^a_b(\eta) f|_{L_m(R^a_b)} + \inf_{\Sigma f^\alpha = \chi_b(\gamma) f} |M^a_b(\eta) \ J^{-\nu}_{b,\alpha} f^\alpha|_{L_m(R^a_b)} \tag{6.2}$$

Using the spaces $H_\gamma(R^a_b)$ as basis spaces we construct the following chains, $\Lambda_\sigma(H_\gamma(R^a_b))$, of Banach spaces

$$H_{\sigma,\gamma}(R_b^a) = H^m_{\sigma,\nu,\gamma,\eta}(R_b^a) = \{ f \in H_\gamma(R_b^a) : \int_{H_\gamma(R_b^a)} |\Delta^\sigma(h) f|^m \frac{dh}{|h|^{3(k(b)-k(a))}} < \infty$$

(6.3)

with the norm

$$|f|_{H_{\sigma,\gamma}(R_b^a)} = |f|_{H_\gamma(R_b^a)} + \{ \int_{H_\gamma(R_b^a)} |\Delta^\sigma(h) f|^m \frac{dh}{|h|^{(k(b)-k(a))3}} \}^{1/m} \tag{6.4}$$

for $\sigma > o$ and $H_{o,\gamma}(R_b^a) = H_\gamma(R_b^a)$. Furthermore, we set

$$\hat{H}^a_{\sigma,\gamma} = \sum_{b \subset a, b \in \bar{A}} \oplus H_{\sigma,\gamma}(R_b^a) \, . \tag{6.5}$$

Besides we will need the following spaces

$$H^a_{\sigma,\gamma} = \bigcap_{b \subset a} H_{\nu_o,\eta_o}(R^b)' \times H_{\sigma,\gamma}(R_b^a) \, . \tag{6.6}$$

The norm in $H^a_{\sigma,\gamma}$ is defined in a standard way.

It is evident from the notation that we are omitting three of the indices defining the spaces. These indices obey the following conditions:

$$\frac{3}{m} < \nu < \frac{1}{2} , \sigma + \nu < \nu_o , \quad \sigma < \frac{1}{2} + \frac{1}{m} \tag{6.7}$$

$$\frac{3}{2} - \frac{3}{m} < \eta < \nu_o - \frac{3}{m} \, .$$

In all other respects the indices are arbitrary, but fixed throughout this paper. The index γ is always supposed to vary in the domain

$$\lambda - (1 - \frac{1}{4k(a)}) \kappa < \gamma < \lambda - \frac{1}{2k(a)} \kappa , \quad \kappa = \min_{c \, d} (\lambda^c - \lambda^d) , \tag{6.8}$$

where λ is a real part of the resolvent parameter (z) which operators considered in these spaces depend on. In order to avoid repeating this condition every time we use the notation $H_{\sigma,\lambda}(R_b^a)$ $(H_{\sigma,\lambda}^a, \hat{H}_{\sigma,\lambda}^a)$ understanding by this any of the spaces $H_{\sigma,\gamma}(R_b^a)$ $(H_{\sigma,\gamma}^a, \hat{H}_{\sigma,\gamma}^a)$ with γ restricted by (6.8).

6.2. The following properties of the spaces, introduced above, may be deduced directly from the definitions:

$$X_{\sigma,\gamma} \quad X_{\sigma,\gamma'}, \; \gamma < \gamma', \; X_{\sigma,\gamma} = H_{\sigma,\gamma}(R_b^a), \; H_{\sigma,\gamma}^a, \; \hat{H}_{\sigma,\gamma}^a \tag{6.9}$$

The spaces $H_{\sigma,\eta,\lambda}(R_b^a)$ are compactly embedded in the spaces $H_{\sigma',\eta',\lambda}(R_b^a)$, $\sigma' < \sigma$, $\eta' < \eta' < \eta$, as follows from their definition and

LEMMA. Suppose that for all α the space $\overset{\alpha}{H}$ is compactly embedded in $\overset{\alpha}{H}'$. Then $H = \Sigma \overset{\alpha}{H}$ is compactly embedded in $H' = \Sigma \overset{\alpha}{H}'$.

PROOF. Set

$$\hat{H} = \Sigma \oplus \overset{\alpha}{H}, \quad \hat{H}' = \Sigma \oplus \overset{\alpha}{H}', \quad J\hat{f} = \Sigma \hat{f}^{\alpha}, \quad f = \{f^{\sigma}\} \in \hat{H} .$$

Any bounded subset M of H is the image under J of some bounded subset $\hat{M}$ of $\hat{H}$. The set $\hat{M}$ is compact in the norm of $\hat{H}'$, and therefore its image $J\hat{M}$ is compact in the norm of H', Q.E.D.

Let now $M \subset R^n$ and X be a Banach space. We define

$$\Lambda_\delta(M,X) = \{f: M \to X: \ |f(m)|_X \leqslant C, \ \ |f(m') - f(m)|_X \leqslant$$

$$\leqslant C|m'-m|^\delta\}, \quad 0 \leqslant \delta \leqslant 1 .$$

Let, furthermore, Ω_b^a be a sphere $\tau(p_b^a) = 1$ in R_b^a. A function $f \in L_m(R_b^a)$ can be considered as a vector function from R^+ into $L_m(\Omega_b^a)$. The following important embedding is true* (see Appendix IV)

$$H_{\sigma,\gamma}(R_b^a)|_{\tau(p_b^a) \geqslant \gamma} \subset \Lambda_\sigma(R^+, L_m(\Omega_b^a)), \ -\sigma \geqslant 0, \ \nu > \frac{1}{m} \tag{6.10}$$

---

*) This embedding in the case $m = 2$ and $\nu > \frac{2}{m} = 1$ was noticed by K. Yajima. It plays an important role in the study of homogeneous equation (6.1). In this study it is required that the solutions of (6.1) belong to $\Lambda_\delta(R^+, \oplus_{b\,a} L_m(\ell_b^a))$, $\delta > \frac{1}{2}$. The embedding used originally by us was, essentially,

$H_{\sigma,\gamma}(R_b^a)|_{\tau(p_b^a) \geqslant \gamma} \subset \Lambda_\delta(R^+, L_m(\Omega_b^a))$, $\delta < \sigma - \frac{1}{m}$. In order to satisfy $\delta > \frac{1}{2}$ it was necessary to consider the scales also in the interval $\frac{1}{2} + \frac{1}{m} \leqslant \sigma \leqslant \frac{1}{2} + \frac{1}{m} + \varepsilon$, $\varepsilon > 0$. This has somewhat complicated the whole study causing us to introduce singular weight functions $(\rho_b^a(\lambda,\varepsilon))^{\alpha}$, in addition to the estimating functions $M_b^a(\eta)$, into the definition of the spaces. Embedding (6.10) enables us to restrict the scales to $0 \leqslant \sigma < \frac{1}{2} + \frac{1}{m}$ so as to avoid this complication. Note that the difference between the two embeddings is that in the original one the $\frac{1}{m}$ price for the passing from $L_m(R^+, L_m(\Omega_b^a))(\equiv L_m(R_b^a))$ to $L_\infty(R^+; L_m(\Omega_b^a))$ is paid from the total σ-smoothness, while in (6.10) (compare Proposition (2.2) of the paper J. Ginibre and M. Moulin, Ann. I.H.P. 21 (1974), 97-145) it is done using the additional (in the variable $(M_{C_i}+M_{C_j})^{-1}(M_{C_i}p_{C_j}-M_{C_j}p_{C_i})$, $(i,j) = \alpha, \{C_i\} = b$) ν-smoothness.

Since the proofs of this paper were still on our desk we have changed the original definition of the spaces, omitting the singular weight functions

(In Appendix IV we prove, for the sake of simplicity, this embedding for the stronger restriction: $\nu > \frac{2}{m}$, which still suits us.)

6.3 Proposition. For any partition $\underline{a}$, the operator $\tilde{R}^a(z)$ defined by (4.33a), (4.38a) may be expressed in the form

$$\tilde{R}^a(z) = P^a(z) + \pi^a(z)\hat{R}^a(z),$$

where the operator $\hat{R}^a(z)$, $z \in \Pi^a$, satisfies the estimates

$$|\hat{R}^a(z)|_{H^a_{\sigma,\lambda} \to \hat{H}^a_{\sigma,\lambda}} < C(1 + |z|)^\delta , \tag{6.11a}$$

$$\int |\Delta^\sigma_\lambda(h)\hat{R}^a(z)|^m_{H^{(a)}_{\sigma,\lambda} \to \hat{H}^a_{\sigma,\lambda}} \frac{dh}{|h|} < C(1 + |z|)^m , \tag{6.12a}$$

$$|(\hat{R}^a(\lambda + i\varepsilon) - \hat{R}^a(\lambda \pm i0))f|_{\hat{H}_{\sigma,\lambda}} + \\ + \int |\Delta^\sigma_\lambda(h)(\hat{R}^a(\lambda + i\varepsilon) - \hat{R}(\lambda \pm i0))f|^m_{\hat{H}_\lambda} \frac{dh}{|h|} \to 0 \tag{6.13a}$$

as $\varepsilon \to \pm 0$, for all $f \in H^{(a)}_{\sigma,\lambda}$. Here and further $\lambda = \text{Re } z$.

6.4. Proof. The proof proceeds by induction. The truth of the proposition for partition in A_n is obvious: $R^a(z) = P^a(z)$, $a \in A_n$. Assuming the proposition true for all b, $b \subset a$, we prove it true for $\underline{a}$. Our argument will be based on investigation of the equation (4.34a) for

---

$\overset{\alpha}{\rho}{}^{\mathbf{a}}_b(\lambda,\varepsilon)$, and have adjusted the exposition throughout the paper. The latter has amounted to crossing out all of these singular functions and dropping some of the estimates.

We are using this opportunity to express our gratitude to K. Yajima for the statement, mentioned above, and for the reading of this paper which has resulted in the corrections of numerous inaccuracies.

the operator $\tilde{R}^a(z)$:

$$\tilde{R}^a(z) + L^a(z)\tilde{R}^a(z) = \tilde{F}^a(z), \tag{6.14}$$

where $L^a(z)$ and $\tilde{F}^a(z)$ are defined by (4.29a) and (4.35a). Henceforth we omit the index a. This will not cause misunderstandings, for we shall not be dealing with any operators or spaces not provided with this index.

Consider the operator $L(z)$. It follows from (4.29a) and (4.30a) that the operator $L(z)$ is a finite linear combination fo monomials of the type

$$\overleftarrow{\prod_{i=1}^{k}} [R_{b_i}(z) V_{\alpha_i}], \tag{6.15}$$

where the partitions b_i and pairs α_i satisfy the conditions

$$b_i \subset \bigcup_1^i u_{\alpha_\ell}, \quad \bigcup_{j=1}^k u_{\alpha_\ell} \not\subseteq b_j;$$

$$\bigcup_{j+1}^i u_{\alpha_\ell} \subseteq b_j \to b_i \subset \bigcup_{j+1}^i u_{\alpha_\ell}, \quad i > j; \quad i,j = 1,\dots,k; \tag{6.16}$$

$$\bigcup_1^k u_{\alpha_\ell} = a. \tag{6.17a}$$

6.5. Let $f \in L_2(R)$. Recall that we are using the notation $f(p_b)$ to emphasize that f is being viewed as vector-valued function from R_b to $L_2(R^b)$. Define operators

$$\bar{R}_b(z) = \tilde{R}_b(z) - P_b(z).$$

It follows from the definitions of the operators $\overline{R}_b(z)$ and $\vec{R}^b(z)$ $(\vec{R}^b(z) = \overline{R}_{b,b}(z))$ that

$$\overline{R}_b(z)f = \vec{R}^b(z - \tau(p_b))f(p_b), \quad f \in L_2(R), \quad b \subset a.$$

Lemma. The operators $\overline{R}_b(z)$ may be expressed as $\overline{R}_b(z) = \pi(z)\hat{R}_b(z)$, where $\hat{R}_b(z) = \pi(z)\hat{R}_b(z)$, where $\hat{R}_b(z)$, $\operatorname{Im} z \neq 0$, satisfy the estimates

$$|\hat{R}_b(z)|_{H^{(b)}_{\sigma,\lambda} \to \hat{H}_{\sigma,\lambda}} \leqslant C(1 + |z|)^{\delta}, \tag{6.18b}$$

$$\int |\Delta^{\sigma}_{z}(h)\hat{R}_b(z)|^{m}_{H^{(b)}_{\sigma,\lambda} \to \hat{H}_{\lambda}} \frac{dh}{|h|} \leqslant C(1 + |z|)^{\delta m}, \tag{6.19b}$$

$$|(\hat{R}_b(\lambda + i\varepsilon) - \hat{R}_b(\lambda \pm i0))f|_{\hat{H}_{\sigma,\lambda}}$$

$$+ \int |\Delta^{\sigma}_{z}(h)(\hat{R}_b(\lambda \pm i\varepsilon) - \hat{R}_b(\lambda \pm i0))f|^{m}_{\hat{H}_{\lambda}} \frac{dh}{|h|} \to 0 \tag{6.20b}$$

as $\varepsilon \to \pm 0$, for all $f \in H^{(b)}_{\sigma,\lambda}$. Here

$$H^{(b)}_{\sigma,\lambda} = \bigcap_{c \subset b} H_{\nu_o,\eta_o}(R^c)' \otimes H_{\sigma,\lambda}(R_c) \subset H_{\sigma,\lambda}$$

Proof. First using the induction hypothesis from the proof of Proposition 6.3 we show (6.18b) - (6.20b) for the auxilary spaces obtained from $H^{(b)}_{\sigma,\gamma}$ and $\hat{H}_{\sigma,\gamma}$ by replacing the estimating functions $M_c(\eta)$ with $M^b_c(\eta)M_b(\eta)$ and restricting the summation over α's .

Then we use the equation

$$\bar{R}_b(z) = -\bar{R}_b(z)L_b'(z) + \bar{F}_b(z) , \qquad (*)$$

where

$$\bar{F}_b(z) = \tilde{F}_b'(z) - P_b(z)\,(E + L_b'(z)) ,$$

implied by (4.37b), to fit this result into the statement of the lemma.

Since the operators $L_b'(z)$ and $\bar{F}_b(z)$ are, in their turn, expressed through $\bar{R}_c(z)$, $c \subset b$, in order to estimate them one has to make assumptions about $\bar{R}_c(z)$, $c \subset b$.

6.7. We prove the lemma by induction on partitions b. If $b \in A_n$ the assertion is trivial: $R_b(z) = 0$, $b \in A_n$. Assuming it is true for all partitions c, $c \subset b$, we prove it for b.

6.8. For technical reasons mentioned above we introduce the following auxiliary spaces: $H_\gamma^{(b)}(R_c)$, $c \subset b \subset a$, is the Banach space obtained from $H_\gamma(R_c)$ by replacing in the definition of the latter, $M_c(\eta)$ by $M_c^b(\eta)M_b(\eta)$ and restricting the summation over the α's by $c_\alpha \subset b$. We let $H_{\sigma,\lambda}^{(b)}(R_c) = \Lambda_\sigma(H_\gamma^{(b)}(R_c))$.

Similarly to (6.5) we define the spaces

$$\hat{H}_{\sigma,\lambda}^{(b)} = \sum_{c \subset b,\, c \in \bar{A}} \oplus\, H_{\sigma,\lambda}^{(b)}(R_c)$$

Besides we introduce also

$$\tilde{H}_{\sigma,\gamma}^{(b)} = \bigcap_{c \subset b} H_{\nu_o,\eta_o}(R^c)' \otimes H_{\sigma,\gamma}^{(b)}(R_c^a)$$

Note that

$$H^{(b)}_{\sigma,\lambda}(R_b) = H_{\sigma,\lambda}(R_b) .$$

The parameters figuring in the definitions of these spaces are subject to the same constraints (6.7).

Now we write down the properties of the spaces defined above which caused the introduction of them.

$$|f|_{H^{(b)}_{\gamma}(R_c)} = \Big|\, |M_b(\eta)|\, f(p_b)|_{H_{\gamma-\tau(p_b)}(R^b_c)} \Big|_{L_m(R_b)} ,$$

$$|f|_{H^{(b)}_{\sigma,\gamma}(R_c)} \leq \Big|\, |M_b(\eta)|\, f(p_b)|_{H_{\sigma,\gamma-\tau(p_b)}(R^b_c)} \Big|_{L_m(R_b)} +$$

$$+ \Big| M_b(\eta) \Big\{ \int \big|\, |\Delta^{\sigma}_{p_b}(h) f(p_b)|^m_{H_{\gamma-\tau(p_b)}(R^b_c)} \frac{dh}{|h|^{3(k(b)-k(a))}} \Big\}^{\frac{1}{m}} \Big|_{L_m(R_b)} , \tag{6.21}$$

$$\Big|\, |M_b(\eta)|\, f(p_b)|^m_{H_{\sigma,\gamma-\tau(p_b)-\delta}(R^b_c)} \Big|_{L_m(R_b)} +$$

$$\Big| M_c(\eta) \Big[\int \big|\, |\Delta^{\sigma}_{p_b}(h) f(p_b)|^m_{H_{\gamma-\tau(p_c)-\delta}(R^b_c)} \frac{dh}{|h|^{(k(b)-k(a))3}} \Big]^{\frac{1}{m}} \Big|_{L_m(R_b)} \leq C|f|_{H^{(b)}_{\sigma,\gamma}(R_c)} , \tag{6.22}$$

where $u_\alpha \subseteq b$, $\sigma > 0$ and $\delta = \frac{1}{4}(k(a)^{-1} - k(b)^{-1})\ \kappa$. The same inequalities are true for other types of spaces defined above.

Note that the first two inequalities follow directly from the definitions of the spaces involved and the last one is based on some general properties of $\Lambda_\alpha^{p,q}$-spaces (in the terminology of [34]) which can be found in [35, pp. 217, 226].

6.9. Lemma. Let $\hat{B}(z)$ be an operator satisfying the estimates (6.11b)-(6.13b) and

$$|D_\lambda \hat{B}(z)|_{H_{\sigma,\eta}(R^b) \to \hat{H}^b_{\nu,\eta}} < C(1 + |\lambda|)^{-1}, \quad \lambda = \mathrm{Re}\, z \lesssim \lambda^b - 1, \tag{6.23}$$

and $\hat{B}'(z)$, the operator defined in $L_2(R)$ by

$$\hat{B}'(z)f = \hat{B}(z - \tau(p_b))f(p_b), \quad f \in L_2(R).$$

Then

$$|\hat{B}'(z)|_{\tilde{H}^{(b)}_{\sigma,\lambda} \to \hat{H}^{(b)}_{\sigma,\lambda}} \lesssim C(1 + |z|)^\delta,$$

$$\int |\Delta_z^\sigma(h)\hat{B}'(z)|^m_{\tilde{H}^{(b)}_{\sigma,\lambda} \to \hat{H}^{(b)}_\lambda} \frac{dh}{|h|} \lesssim C(1 + |z|)^{\delta m},$$

$$|\hat{B}'(\lambda + i\varepsilon) - \hat{B}'(\lambda \pm i0))f|_{\hat{H}^{(b)}_{\sigma,\lambda}}$$

$$+ \int |\Delta_z^\sigma(h)(\hat{B}'(\lambda + i\varepsilon) - \hat{B}'(\lambda \pm i0))f|^m_{\hat{H}^{(b)}_\lambda} \frac{dh}{|h|} \to 0$$

as $\varepsilon \to \pm 0$, for all $f \in \tilde{H}^{(b)}_{\sigma,\lambda}$. Here $\hat{H}_{\nu,\eta} = \sum_{c \subset d} \oplus H_{\nu,\eta}(R_c^b)$.

Proof. Since the estimates of the statement of the lemma are established in the same way, we shall prove only one of them -- the first.

Let $f \in \tilde{H}^{(b)}_{\sigma,\gamma}$ and $\sigma > 0$ be taken for a definity. Using (6.21) we obtain

$$|\hat{B}'(z)f|_{\hat{H}^{(b)}_{\sigma,\gamma}} \le \Big|M_b(\eta)\,|\hat{B}(z-\tau(p_b))f(p_b)|_{\hat{H}^b_{\sigma,\gamma-\tau(p_b)}}\Big|_{L_m(R_b)}$$

$$+ \Big|M_b(\eta)\Big[\int |\Delta^{\sigma}_{p_b}(h)\hat{B}(z-\tau(p_b))f(p_b)|^m_{\hat{H}^b_{\gamma-\tau(p_b)}}\frac{dh}{|h|}\Big]^{1/m}\Big|_{L_m(R_b)}$$

The estimate for the first term on the right follows from estimate (6.11b) for $\hat{B}(z)$ and inequality (6.22). Now we estimate the second term on the right of this inequality. We have

$$|\Delta^{\sigma}_{p_b}(h)\hat{B}(z-\tau(p_b))f(p_b)|_{\hat{H}_{\gamma-\tau(p_b)}}$$

$$\le |(\Delta^{\sigma}_{p_b}(h)\hat{B}(z-\tau(p_b)))f(p_b)|_{\hat{H}^b_{\sigma-\tau(p_b)}} + |\hat{B}(z-\tau(p_b+h))\Delta^{\sigma}_{p_b}(h)f(p_b)|_{\hat{H}^b_{\gamma-\tau(p_b)}}$$

$$\le |\Delta^{\sigma}_{p_b}(h)\hat{B}(z-\tau(p_b))|_{H^b_{\sigma,\gamma-\tau(p_b)+\delta}\to\hat{H}^b_{\gamma-\tau(p_b)}}\,|f(p_b)|_{H^b_{\sigma,\gamma-\tau(p_b)+\delta}}$$

$$+ |\hat{B}(z-\tau(p_b+h))|_{H^b_{\gamma-\tau(p_b)+\delta}\to\hat{H}^b_{\gamma-\tau(p_b)}}\,|\Delta^{\sigma}_{p_b}(h)f(p_b)|_{H^b_{\gamma-\tau(p_b)+\delta}} \qquad (**)$$

where $\delta = \frac{1}{4}(k(a)^{-1}-k(b)^{-1})\kappa$. We estimate the first term on the right of (**). Let $|h| < 1$ and $\tau(p_b) > |\lambda| - \lambda^b + 5$. Using the estimate (6.23) for $\hat{B}(z)$, we obtain

$$|\hat{B}(z-\tau(p_b+h)) - \hat{B}(z-\tau(p_b))|_{H^b_{\sigma,\gamma-\tau(p_b)+\delta}\to\hat{H}^b_{\gamma-\tau(p_b)}}$$

$$< C|D_\lambda \hat{B}(z - \tau(p_b + \bar{h})|_{H_{\sigma,\eta}(R^b) \to \hat{H}^b_{\nu,\eta}} |\tau(p_b + h) - \tau(p_b)|$$

$$< C|h| \sup_{\mathrm{Re} w < \lambda^b - 5} (1 + |\mathrm{Re} w|) | D_{\mathrm{Re}\, w} \hat{B}(w)|_{H_{\sigma,\eta}(R^b) \to \hat{H}^b_{\nu,\eta}} .$$

Now let $\tau(p_b) < |\lambda| - \lambda^b + 5$ (we are assuming that $|h| < 1$; otherwise the estimate is trivial).

We distinguish between two cases: $|\tau(p_b + h) - \tau(p_b)| < 10h^2$ and $|\tau(p_b + h) - \tau(p_b)| > 10h^2$. In the first case we use the estimate

$$|\hat{B}(z + h) - \hat{B}(z)|_{H^b_{\sigma,\gamma+\delta} \to \hat{H}^b_\gamma} < C|h|^{\sigma'}, \quad \sigma' < \sigma,$$

which follows from the estimate

$$\int |\hat{B}(z + h) - \hat{B}(z)|^m_{H^b_{\sigma,\gamma+\delta} \to \hat{H}^b_\gamma} \frac{dh}{|h|^{1+\sigma m}} < C.$$

In the second case we change variables: $h \to \ell = \tau(p_b) - \tau(p_b + h)$ and use the estimates

$$\frac{dh}{d\ell} < C \frac{|h|}{|\ell|}, \quad |\ell| < C|h|(1 + |\lambda|)^{1/2} .$$

These operations yield the following estimate for the first term on the right of (**):

$$(1 + |z|)^{\sigma/2} |M_b(\eta)|f(p_b)|_{H^b_{\sigma,\gamma-\tau(p_b)+\delta}} \times$$

$$\times \{ \int [|\Delta^\sigma(\ell)\hat{B}(z - \tau(p_b))|^m_{H^b_{\sigma,\gamma-\tau(p_b)+\delta} \to \hat{H}^b_{\gamma-\tau(p_b)}} \frac{d\ell}{|\ell|}]^{1/m}$$

$$+ \sup_{\operatorname{Re} w < \lambda^b - 5} (1 + |\operatorname{Re} w|) \| D_{\operatorname{Re} w} \hat{B}(w) \|_{H_{\sigma,\eta}(R^b) \to \hat{H}^b_{\nu,\eta}} \} \|_{L_m(R_b)}$$

$$\lesssim C(1 + |z|)^{\frac{\sigma}{2}+\delta} \| M_b(\eta) \| f(p_b) \|_{H^b_{\sigma,\gamma-\tau(p_b)+\delta}} \|_{L_m(R_b)}$$

$$\lesssim C(1 + |z|)^{\delta+\frac{\sigma}{2}} \|f\|_{\tilde{H}^{(b)}_{\sigma,\gamma}} .$$

Proceeding to the second term on the right of (**), we have the estimate

$$(1 + |z|)^{\delta} \| M_b(\eta) [\int \| \Delta^{\sigma}_{p_b}(h) f(p_b) \|^m_{H^b_{\gamma-\tau(p_b)+\delta}} \frac{dh}{|h|}]^{1/m} \|_{L_m(R_b)}$$

$$\lesssim (1 + |z|)^{\delta} \|f\|_{\tilde{H}^{(b)}_{\sigma,\gamma}} .$$

To obtain the last terms in these two inequalities, we have used inequality (6.22).

This completes the proof of Lemma 6.9.

REMARK. Estimates of type (6.23) for $\hat{R}^b(z)$ can be easily obtained in the same way as the corresponding estimates of Section 5.

6.10. In order to use Equation (*) one has to estimate $L_b'(z)$ and $\overline{F}_b(z)$. The operator $L_b'(z)$ is a finite linear combination -- with the same coefficients as $L_b(z)$ -- of monomials

$$\overset{\curvearrowright}{\prod_{i=1}^{k}} [V_{\alpha_i} R_{c_i}(z)] , \tag{6.24}$$

with α_i and c_i satisfying conditions (6.16), (6.17b).

LEMMA. Let $A(z)$ have the form of (6.24) and assume that conditions (6.16), (6.17b) are satisfied. Then $A(z)$, $z \in \Pi^b$, satisfies the estimates

$$|A(z)|_{H^{(b)}_{\sigma,\lambda} \to \tilde{H}^{(b)}_{\sigma,\lambda}} < C(1 + |z|)^{\delta} ,$$

$$\int |\Delta^{\sigma}_{\lambda}(h) A(z)|^{m}_{H^{(b)}_{\sigma,\lambda} \to \tilde{H}^{(b)}_{\lambda}} \frac{dh}{|h|} < C(1 + |z|)^{\delta m} ,$$

$$|(A(\lambda + i\varepsilon) - A(\lambda \pm i0))f|_{\tilde{H}^{(b)}_{\sigma,\lambda}}$$

$$+ \int |\Delta^{\sigma}_{\lambda}(h)(A(\lambda + i\varepsilon) - A(\lambda \pm i0))f|^{m}_{\tilde{H}^{(b)}_{\lambda}} \frac{dh}{|h|} \to 0$$

as $\varepsilon \to \pm 0$, for all $f \in H^{(b)}_{\sigma,\lambda}$.

Proof of Lemma 6.10. By the induction hypothesis (beginning of proof of Lemma 6.5), the operators $\bar{R}_c(z)$, $c \subset b$, may be written as

$$\bar{R}_c(z) = \pi(z)\hat{R}_c(z), \tag{6.25}$$

where $\hat{R}_c(z)$ satisfies the estimates (6.18c)--(6.20c) and is strongly continuous in $\text{Im } z$. Let

$$\hat{P}_v f = \{(\hat{P}_v f)_g = \delta_{v,g} P_v f,\ g \subset a\},\quad f \in L_2(R),$$

where

$$(P_v f)(p_v) = (\psi^v, f(p_v)) \quad \text{if} \quad v \in \bar{A} \quad \text{and} \quad P_v = 0 \quad \text{if} \quad v \notin \bar{A}$$

Set

$$\hat{R}'_c(z) = \sum_{v \ c} \{\hat{P}_v + \hat{R}_v(z)\},\quad c \subset b. \tag{6.27}$$

It follows from (4.38c), (4.40c), (6.25) and (6.27) that

$$R_c(z) = \pi(z)\hat{R}'_c(z). \tag{6.28}$$

Inserting (6.28) in the formula for $A(z)$, we obtain

$$A(z) = \overset{\curvearrowright}{\prod_{i=1}^{k}} [V_{\alpha_i} \pi(z) \hat{R}'_{c_i}(z)].$$

6.11. Thus, in order to establish estimates for $A(z)$, we have to study the operators V_α and $R_o(z)V_\alpha$ for different pairs α. This we now proceed to do.

Set

$$R_{T_c}(z)\Phi = (\lambda^c + \tau(p_c) - z)^{-1}\,\Phi(p), \quad c \in \bar{A} .$$

Lemma . Let $u_\alpha \not\subseteq c$, $f \in L_m(R_c)$ and suppose that $\phi(p^c)$ satisfies the estimate (5.24). Then the function $V_\alpha R_{T_c}(z)\,\phi\cdot f$, $\operatorname{Im} z \neq 0$, obeys the estimates

$$|V_\alpha R_{T_c}(z)\phi f|_{H^{(e)}_{\sigma,\lambda}(R)} < C|V_\alpha|_o\,|\phi|_c\,|f|_{H^{(d)}_{\sigma,\lambda}(R_c)}(1 + |z|)^\delta,$$

$$\{\int |\Delta^\sigma_\lambda(h)V_\alpha R_{T_c}(z)\phi f|_{H^{(e)}_\lambda(R)}\frac{dh}{h}\}^{1/m} < C|f|_{H^{(d)}_{\sigma,\lambda}(R_c)}(1 + |z|)^\delta,$$

$$|V_\alpha R_{T_c}(\lambda + i\varepsilon)\phi f - V_\alpha R_{T_c}(\lambda \pm i0)\phi f|_{H^{(e)}_{\sigma,\lambda}(R)}$$

$$+ \int \Delta^\sigma_\lambda(h)\,(V_\alpha R_{T_c}(\lambda + i\varepsilon)\phi f - V_\alpha R_{T_c}(\lambda \pm i0)\phi f)|^m_{H^{(e)}_\lambda(R)}\frac{dh}{h} \to 0$$

as $\varepsilon \to \pm 0$ for all $f \in H^{(d)}_{\sigma,\lambda}(R)$. Here $e = dUu_\alpha$.

The proof of Lemma 6.11 is given in Appendix II.

6.12. Now consider the operator $R_o(z)V_\alpha$. Let the functions ϕ^c $c \subset a$, satisfy (5.14), and $f_c \in L_2(R_c)$, $c \subset a$. We have

$$R_o(z)V_\alpha \sum_{c\subset a} R_{T_c}(z)\phi^c\cdot f_c$$

$$= \sum_{c\supseteq u_\alpha} R_o(z)R_{T_c}(z)V_\alpha^c\phi^c\cdot f_c + \sum_{c\not\supseteq u_\alpha} R_o(z)V_\alpha R_{T_c}(z)\phi^c\cdot f_c.$$

The terms in the first sum on the right may be transformed thus:

$$R_o(z)R_{T_c}(z)V_\alpha^c\phi^c\cdot f_c = R_{T_c}(z)\tilde{\phi}^c\cdot f_c - R_o(z)\tilde{\phi}^c\cdot f_c ,$$

where

$$\tilde{\phi}^c(p^c) = (\tau(p^c) - \lambda^c)^{-1}(V_\alpha^c\phi^c)(p^c).$$

This gives

$$R_o(z)V_\alpha \sum R_{T_c}(z)\phi^c f_c$$

$$= R_o(z)(-\sum_{c\supseteq u_\alpha} \tilde{\phi}^c f_c + \sum_{c\not\supseteq u_\alpha} V_\alpha R_{T_c}(z)\phi^c f_c) + \sum_{c\supseteq u_\alpha} R_{T_c}(z)\tilde{\phi}^c f_c.$$

It follows from (6.27) that the functions $\tilde{\phi}^c(p^c)$, $c \subset a$, obey the estimate (5.24) with the constant $C|v_\alpha|_o|\phi^c|_c$. Hence

$$|\tilde{\phi}^c f_c|_{H^{(d)}_{\sigma,\lambda}(R)} \leq C|v|_o|\phi^c|_c \, |f_c|_{H_{\sigma,\lambda}(R_c)} \qquad (|f_c|_{H^{(d)}_{\sigma,\lambda}(R_c)}),$$

where $c \supseteq d \cup u_\alpha (u_\alpha \subseteq c \subseteq d)$.

6.13. Now we proceed to the estimation of $A(z)$. In order to avoid the lengthy expressions repeating in the sequel we introduce the new notation:

$$\mathscr{B}_\delta(X_{\sigma,\lambda}, Y_{\sigma,\lambda}) = \{A(z) : |A(z)|_{X_{\sigma,\lambda}\to Y_{\sigma,\lambda}} \leq C(1 + |z|)^\delta ,$$

$$\int |\Delta_z^\sigma(h)A(z)|^m_{X_{\sigma,\lambda}\to Y_{\sigma,\lambda}} \frac{dh}{|h|} \leq C(1 + |z|)^{\delta m} ,$$

$$|(A(\lambda + i\varepsilon) - A(\lambda \pm i0))f|_{Y_{\sigma,\lambda}} +$$

$$+ \int |\Delta_\lambda^\sigma (A(\lambda + i\varepsilon) - A(\lambda \pm i0))f|^m_{Y_{\sigma,\lambda}} \frac{dh}{|h|} \to 0 \quad (\varepsilon \to 0)\ \forall f \in X_{\sigma,\lambda},\ \lambda = \mathrm{Re}z\}$$

6.14. Now we consider monomial (6.24) with conditions (6.16) and (6.17b). It can be decomposed, using Eqns (*), into a sum of monomials, each one of which is a product of the factors of the forms

$$\overset{\curvearrowright}{\prod_{i=s}^{t}} [V_{\alpha_i} R_{c_i}(z)] , \tag{6.29}$$

satisfying

$$c_i \subset \bigcup_{i+1}^{t} u_{\alpha_\ell} , \quad \bigcup_{s+1}^{t} u_{\alpha_\ell} \subseteq c_t, \quad u_{\alpha_s} \nsubseteq c_t \tag{6.30}$$

We will prove for each such factor that

$$\overset{\curvearrowright}{\prod_{i=s}^{t}} [V_{\alpha_i} R_{c_i}(z)] \in \mathscr{B}_\delta(\tilde{H}^{(d)}_{\sigma,\lambda}, \tilde{H}^{(e)}_{\sigma,\lambda}) \cap \mathscr{B}_\delta(H^{(b)}_{\sigma,\lambda}, \tilde{H}^{(f)}_{\sigma,\lambda}), f = \bigcup_{i=s}^{t} u_{\alpha_1}, \ e = d \cup f, \ d \supseteq c_t \tag{6.31}$$

Then a combination of these results gives us the desired result for the whole product (6.24).

6.15. In order to simplify slightly expressions we impose the additional condition

$$u_{\alpha_\ell} \nsubseteq c_\ell, \qquad \ell = s,\ldots,t-1, \tag{6.32}$$

for every factor of the form (6.29), (6.30).

In the general case the considerations are changed not essentially, only the range of the summation in (6.33) gets more complicated. Besides this, the estimates for the general case can be reduced to the estimates for the cases with this additional condition.

Lemma. The monomials $\prod_{i=j}^{t} [V_{\alpha_i} R_{c_i}(z)]$, $j = t, t-1, \ldots, s$,

with conditions (6.30) and (6.32) obey the estimates

$$\overset{\rightarrow}{\prod_{i=j}^{t}} [V_{\alpha_i} R_{c_i}(z)] \in \mathscr{B}_\delta(H_{\sigma,\lambda}^{(d)}, H_{\sigma,\lambda}^{(e_j)} + \sum_{d_j \subseteq v \subseteq c_t} \phi_j^v(z - \tau(p_v)) R_{T_v}(z) H_{\sigma,\lambda}^{(e)}(R_v)), \tag{6.33j}$$

and the same with $\tilde{H}_{\sigma,\lambda}^{(d)}$ and $\tilde{H}_{\sigma,\lambda}^{(e_j)}$ replaced by $H_{\sigma,\lambda}^{(b)}$ and $\tilde{H}_{\sigma,\lambda}^{(d_j)}$,

where $d \supseteq c_t$ or $d \in A_n$, $e = d \cup c_t$, $e_j = u_j \cup e$, $d_j = \bigcup_j^t u_i$ and

$$\phi_j^v(z) = \overset{\rightarrow}{\prod_{i=j}^{t-1}} [V_{\alpha_i}^v \pi^v(\lambda^v) \hat{R}'_{c_i,v}(z)] V_{\alpha_t}^v \psi^v \quad (d_j \subseteq v \subseteq c_t) \tag{6.34}$$

Remark. (5.33s) coincides with desired estimate (6.31).

6.16. Proof. Since the denominators contained in $\pi^v(\lambda^v)$ are not singular one can prove easily, using only the induction hypothesis about $\hat{R}_{c,v}(z)$, $c \subset v$, the following estimates for (6.34)

$$|\phi_j^v(z)|_{H_{\sigma,\lambda}(R^v)} + \int |\Delta_z^\sigma(h) \phi_j^v(z)|^m_{H_\lambda(R^v)} \frac{dh}{|h|} < C$$

and

$$|D_z^k \phi_j^v(z)|_{H_{\nu_o,\eta_o}(R^v)} < C \quad \forall k, \text{ when } z \notin [\mu^\sigma, \infty) .$$

6.17. We prove (6.33j) by an induction on j. Let (6.33j) be valid for all $j \geqslant k+1$; prove it for $j = k$.

Consider the function $\overset{\rightarrow}{\prod_{i=k+1}^{t}} [V_{\alpha_i} R_{c_i}(z)] f$ for some $f \in H_{\sigma,\lambda}^{(d)}$. It can be decomposed in accordance with the right side of (6.33 k+1). Let $g(z)$ be the $H_{\sigma,\lambda}^{(e_{k+1})}$ - component of $\overset{\rightarrow}{\prod_{i=k+1}^{t}} [V_{\alpha_i} R_{c_i}(z)] f$. We begin with the estimation of the function $V_{\alpha_k} R_{c_k}(z)\, \bar{g}(z)$.

Note first of all that the spaces $\tilde{H}^{(e)}_{\sigma,\lambda}$ are designed in such a way that

$$\hat{P}_v \in \mathscr{L}(\tilde{H}^{(e_{k+1})}_{\sigma,\lambda} , \tilde{H}^{(e_{k+1})}_{\sigma\ \lambda}) , \qquad v \subset e_{k+1} ,$$

where the operator $\hat{P}_v$ is defined by (6.25) . These relations, the induction hypothesis about $\hat{R}_v(z)$, $v \subset b$, Eqn (6.28) and Lemma 6.12 imply that

$$V_{\alpha_k} R_{c_k}(z) \in \mathscr{B}_\delta (\tilde{H}^{(e_{k+1})}_{\sigma,\lambda} , \hat{H}^{(e_k)}_{\sigma,\lambda}) \tag{*}$$

LEMMA. The function $g(z) = V_{\alpha_k} R_{c_k}(z)\, \bar{g}(z)$ satisfies the estimates

$$|(\phi,g(z))|_{H^{(e_{k+1})}_{\sigma,\lambda}(R_v)} +$$

$$+ \int |\Delta^\sigma_z(h)(\phi,g(z))|^m_{H^{(e_{k+1})}_\lambda(R_v)} \frac{dh}{|h|} \lesssim C\, |\phi|_{H_{\nu_o,\eta_o}(R^v)} , \qquad v \subset c_k \tag{6.35}$$

The proof of the lemma is given in the end of this section, § 6.38.

Using this lemma and (*) we obtain

$$|V_{\alpha_k} R_{c_k}(z) g(z)|_{\tilde{H}^{(e_k)}_{\sigma,\lambda}} +$$

$$+ \int \left| \Delta_z^\sigma(h)\, V_{\alpha_k} R_{c_k}(z) g(z) \right|_{\underset{H_\lambda}{\sim}^m (e_k)} \frac{dh}{|h|} \leq C$$

6.18. The estimates for the operator $V_{\alpha_k} R_{c_k}(z)$ with the second term in the right side of (6.33 k+1) are lengthy and simple and we will not write them down here. We only explain briefly how to obtain them. First of all we note that

$$V_{\alpha_k} R_{c_k}(z) \sum_{d_{k+1} \subset v \subset c_t} \phi_{k+1}^v(z - \tau(\cdot)) R_{T_v}(z) f_v =$$

$$= V_{\alpha_k} \sum_{d_{k+1} \subset v \subset c_t} \pi(z)\ (\hat{R}'_{c_k,v}(z - \tau(\cdot)) \phi_{k+1}^v(z - \tau(\cdot))) R_{T_v}(z) f$$

Then we use the identity

$$(\lambda^v + \tau(p_v) - z)^{-1} (\lambda^w + \tau(p_w) - z)^{-1} =$$

$$= (\lambda^w - \lambda^v + \tau(p_w^v))^{-1} \{ (\lambda^v + \tau(p_v) - z)^{-1} - (\lambda^w + \tau(p_w) - z)^{-1} \},$$

$$w \subset v,$$

in order to separate the singularities of $\pi(z)$ and $R_{T_v}(z)$ and pick out the summand

$$\sum_{d_k \subset v \subset c_t} V_{\alpha_k}^v\ \pi^v(\lambda^v) \hat{R}'_{c_k,v}(z - \tau(\cdot)) \phi_{k+1}^v(z - \tau(\cdot)) R_{T_v}(z) f$$

$$= \sum_{d_k \subset v \subset c_t} \phi_k^v(z - \tau(\cdot)) R_{T_v}(z) f$$

In order to estimate the remainders one has to estimate $\hat{R}'_{c_k,v}(z) \phi_{k+1}^v(z)$ and to use the trivial modification of Lemma 6.12.

This completes the proof of Lemma 6.15, which, in correspondence with

Remark 6.15, gives the proof of estimate (6.31). Lemma 6.10 is proved.

6.19. Expressing $L_b'(z)$ through the monomials of the form (6.24) with restrictions (6.16) and (6.17b) and applying to each of these monomials Lemma 6.10 proved above we get

Corollary. The operator $L_b'(z)$ satisfies estimates of the statement of Lemma 6.10.

6.20. Now we proceed to the consideration of the operator $\overline{F}_b(z)$. It can be transformed as follows

$$\overline{F}_b(z) = F_b'(z) - G_b(z)(E + L_b'(z)) - P_b(z)(H_o - zE)F_b'(z)$$

$$= (E - P_b)F_b'(z) - G_b(z)(E + L_b'(z))$$

$$= M_b'(z) - G_b(z)L_b'(z) - \overline{P}_b F_b(z) \quad ,$$

where $G_b(z) = \sum_{c \subset b} R_c(z)$ and

$$(\overline{P}_b f)(p) = \psi^b(p^b)(\psi^b, f(p_b)), \quad b \in \bar{A} \setminus A_n,$$

$$\overline{P}_b = 0, \quad b \notin \bar{A} \setminus A_n \ .$$

Using similar reasoning as in the proof of Lemma 6.10, one can prove

Lemma. The operator $\overline{F}_b(z)$ may be expressed in the form

$$\overline{F}_b(z) = \pi(z)\hat{F}_b(z) \ ,$$

where the operator $\hat{F}_b(z)$ satisfies estimates (6.19b) - (6.21b).

6.21. The statement of Lemma 6.5 follows from Lemmas 6.9 and 6.20 and Corollary 6.19 (see also Remark 6.9) applied to the equation

$$\hat{R}_b(z) = -\hat{R}_b(z)L_b'(z) + \hat{F}_b(z)$$

The proof is complete.

6.22. We can now extend definition (6.26) to all b, $b \subset a$;

$$\hat{R}'_b(z) = \sum_{c \subseteq b} \{\hat{P}_c + \hat{R}_c(z)\}, \qquad b \subset a .$$

Let $\hat{L}(z)$ be the operator obtained from $L(z)$, expressed as the linear combination of (6.15), by substituting

$$\overset{k}{\underset{i=1}{\curvearrowleft\!\!\prod}} [\hat{R}'_{b_i}(z) V_{\alpha_i} \pi(z)] \tag{6.36}$$

for (6.15). It follows from definition of $\hat{L}(z)$ that

$$L(z)\pi(z) = \pi(z)\hat{L}(z).$$

Lemma. The operator $\hat{L}(z)$, $z \in \Pi$, defined above satisfies the estimates

$$|\hat{L}(z)|_{\hat{H}_{\sigma,\lambda} \to \hat{H}_{\sigma,\lambda}} \lesssim C(1 + |z|)^{\delta},$$

$$\int |\Delta_z^{\sigma}(h)\hat{L}(z)|^m_{\hat{H}_{\sigma,\lambda} \to \hat{H}_{\lambda}} \frac{dh}{|h|} \lesssim C(1 + |z|)^{\delta m},$$

$$|(\hat{L}(\lambda + i\varepsilon) - \hat{L}(\lambda \pm i0))\hat{f}|_{\hat{H}_{\sigma,\lambda}} + \int |\Delta_\lambda^{\sigma}(h)(\hat{L}(\lambda + i\varepsilon) - \hat{L}(\lambda \pm i0))\hat{f}|^m_{\hat{H}_\lambda} \times \frac{dh}{|h|} \to 0$$

as $\varepsilon \to 0$, for all $\hat{f} \in \hat{H}_{\sigma,\lambda}$.

6.23. Proof. In order to obtain the statement of the lemma one applies successively Lemma 6.5 and the simple modifications of Lemmas 6.12 and 6.15 to the factors in every monomial of the form (6.36) which enters into the expression for $\hat{L}(z)$ and take into account restrictions (6.16) and (6.17a). The considerations used for this are exactly the same as ones used in the proof of Lemma 6.10, only Lemma 6.5 is used instead of the induction hypothesis from its proof. We repeat here shortly the main points of these considerations adapting them to the case in hand.

a. We consider monomial (6.36) with conditions (6.16) and (6.17a) on the indices b's and α's. It can be decomposed into a product of factors of the form

$$\overleftarrow{\prod_{i=s}^{t}} [\hat{R}'_{b_i}(z) V_{\alpha_i} \pi(z)] \tag{6.37}$$

with the restrictions

$$b_i \subset \bigcup_{\ell=s}^{i} u_{\alpha_\ell}, \quad \bigcup_{\ell=s}^{t-1} u_{\alpha_\ell} \subset b_{s-1} \quad (s \leq i \leq t-1), \quad u_{\alpha_t} \subseteq b_{s-1} \tag{6.38}$$

The aim is to prove that

$$(6.37), (6.38) \in \mathscr{B}_\delta\Big(\bigoplus_{c \subseteq b_{s-1}} H_{\sigma,\lambda}(R_c), \bigoplus_{c \subset b_t} H_{\sigma,\lambda}(R_c)\Big) \tag{6.39}$$

Then the statement of Lemma 6.20 is obtained by the successive application of this result.

b. In order to simplify slightly subsequent expressions we assume that every factor satisfies, in addition to (6.38), also the restriction

$$u_{\alpha_{\ell+1}} \nsubseteq b_\ell, \quad s+1 \leq \ell \leq t-1 \tag{6.38'}$$

The justification of this restriction is the same as of the corresponding restrictions of §6.15.

c. Lemma. The monomials $\overleftarrow{\prod_{i=s}^{j}} [\hat{R}'_{b_i}(z) V_{\alpha_i} \pi(z)]$ $(j = s,\dots,t)$ with conditions (6.38) and (6.38') on the b_i's and the α_i's obey the estimates

$$\overleftarrow{\prod_{i=s}^{j}} [\hat{R}'_{b_i}(z) V_{\alpha_i} \pi(z)] \in \mathscr{B}_\delta\Big(\bigoplus_{c \subseteq b_{s-1}} H_{\sigma,\lambda}(R_c), \bigoplus_{c \subseteq b_j} H_{\sigma,\lambda}(R_c) +$$

$$+ \sum_{d_j \subseteq v \subseteq b_{s-1}} \hat{\phi}^v_j(z - \tau(\cdot)) R_{T_v}(z)\, H_{\sigma,\lambda}(R_v)\Big), \tag{6.40 j}$$

where $d_j = \bigcup_{\ell=s}^{j} u_{\alpha_\ell}$ and

$$\hat{\phi}_j^v(z) = \overleftarrow{\prod_{i=s+1}^{j}} [\hat{R}'_{b_i,v}(z) V^v_{\alpha_i} \pi^v(\lambda^v)] \hat{R}'_{b_s,v}(z) V^v_{\alpha_s} \psi^v$$

Remark. (6.40t) coincides with (6.39).

d. The proof of this lemma is given by induction: (6.40j) is assumed true for all $j \leqslant k-1$ and is proved for $j = k$. In order to prove that $\hat{R}'_{b_k}(z) V_{\alpha_k} \pi(z) g(z)$, where $\hat{g}(z) = \overleftarrow{\prod_{i=s}^{k+1}} [\hat{R}'_{b_i}(z) V_{\alpha_i} \pi(z)] \hat{f}$, belongs for arbitrary $\hat{f} \in \bigoplus_{c \subset b_j} H_{\sigma,\lambda}(R_c)$ to the space standing in the right side of (6.40k) one uses Lemma 6.5 and the simple modifications of Lemmas 6.12 and 6.15.

6.24. Now we proceed to the proof of the compactness of the boundary values of $\hat{L}^3(z)$ at R in the scale $\hat{H}_{\sigma,\lambda}$, $0 \leqslant \sigma < \frac{1}{2} + \frac{1}{m}$, $\frac{3}{2} < \eta < \eta_o - \frac{3}{m}$, of Banach spaces. Here $\lambda = \mathrm{Re}\, z$, the restrictions on the other indices are given in (6.7).

Let $\widehat{\mathfrak{A}}^b_n$ ($\mathfrak{A}^b_n$) denote the set of operators which are products of n factors of type (6.15) ((6.24)) satisfying conditions (6.16) and (6.17b).

Lemma. (i) The operators from $\widehat{\mathfrak{A}}^a_n$ and $\mathfrak{A}^a_n$, $n = 1,2,\ldots,$ are compact in $\hat{H}_{\sigma,\lambda}$ and $H_{\sigma,\lambda}$, respectively, for all $z \in \mathbb{C} \setminus [\mu^a,\infty)$.

(ii) The operators from $\widehat{\mathfrak{A}}^a_3$ and $\mathfrak{A}^a_3$ converge in the norm of $\mathscr{L}(\hat{H}_{\sigma,\lambda}, \hat{H}_{\sigma,\lambda})$ and $\mathscr{L}(H_{\sigma,\lambda}, H_{\sigma,\lambda'})$, respectively, as $\mathrm{Im}\, z \to \pm 0$, and therefore they have compact boundary values at the real axis.

Proof. The first part of the lemma is proved in the same way as Lemma 5.2 of Sec. 5. Thus we proceed directly to the proof of the second part.

For each operator in $\widehat{\mathfrak{A}}^a_3/\mathfrak{A}^a_3$ we shall construct a sequence of compact and continuous in $z \in \Pi^a$ in the operator norm operators converging to it in norm. The compactness of operators in $\widehat{\mathfrak{A}}^a_3/\mathfrak{A}^a_3$ for $z \in \Pi^a$ will then follow from the fact that the set of compact operators in a Banach space is closed in the uniform operator topology.

6.25 Lemma. Let the operator $\hat{B}(z)$ satisfy estimates (6.23) and (6.11b)--(6.13b). Then there exists a sequence of operators $\hat{B}_n(z)$ with kernels in the space $\sum_{d\subset b, d\in\bar{A}} \oplus C^{\infty}[\overline{C^{\pm}}, \quad S(R_d^b \times R^b)]$ such that

$$|\hat{B}_n(z)f - \hat{B}(z)f|_{\hat{H}^b_{\sigma,\lambda}} \to 0 \quad \text{as} \quad n \to \infty, \tag{6.41}$$

$$\int |\Delta_\lambda^\sigma(h)(\hat{B}_n(z) - \hat{B}(z))f|^m_{\hat{H}^b_\lambda} \frac{dh}{|h|} \to 0 \quad \text{as} \quad n \to \infty, \tag{6.42}$$

for all $f \in H^b_{\sigma,\lambda}$ and $\operatorname{Im} z \geq 0$ $(\operatorname{Im} z \leq 0)$,

$$\sup_{\lambda<\lambda^b-1} (1+|\lambda|)|D_\lambda^k(\hat{B}_n(z) - \hat{B}(z))f|_{\hat{H}^b_{\nu,\eta}} \to 0 \quad \text{as} \quad n \to \infty, \tag{6.43}$$

for all $f \in H_{\sigma,\eta}(R^b)$.

Proof. To prove this lemma we shall simply construct the approximating operators $B_n(z)$ explicitly. Let T_δ, $\hat{T}_\delta$ be standard smoothing operators in the spaces $L_2(R^b)$, $\sum_{c\subset b, c\in\bar{A}} \oplus L_2(R_c^b)$, respectively, and Φ_δ, $\hat{\Phi}_\delta$ the multiplication operators defined by

$$(\Phi_\delta f)(p^b) = \Phi(\delta p^b)f(p^b),$$

$$\hat{\Phi}_\delta \hat{f} = \{\Phi_c(\delta p_c^b)f_c(p_c^b),\ c\subset b,\ c\in\bar{A}\},\ \hat{f} = \{f_c(p_c^b),\ c\subset b,\ c\in\bar{A}\},$$

where

$$\Phi \in S(R^b),\quad \Phi(p^b)\equiv 1,\quad |p^b|\leq 1;\quad \Phi_c \in S(R_c^b),\quad \Phi_c(p_c^b)\equiv 1,\quad |p_c^b|\leq 1,\quad c\subset b.$$

Let ε_n, δ_n $(n = 1,2,\ldots.)$ be sequence of numbers converging to 0 as $n\to\infty$, such that $\varepsilon_n\cdot\operatorname{Im} z > 0$. Define operators by

$$\hat{B}_n(z) = \hat{\Phi}_{\delta_n}\hat{T}_{\delta_n}\hat{B}(z+i\varepsilon_n)T_{\delta_n}\Phi_{\delta_n},\ n=1,2,\ldots\ .$$

It is easy to see that these operators meet all the requirements of the lemma, and so the proof is complete.

6.26. The next lemma enables us to replace strong convergence by uniform convergence.

Lemma. Let X and Y be Banach spaces, K a compact operator in X, A_n and A bounded operators from X to Y. Assume that A_n converges strongly to A as $n \to \infty$. Then the operators A_nK converge to A in the norm of $\mathscr{L}(X.Y)$.

Proof of Lemma 6.26. For any $\varepsilon > 0$, we can express K as a sum $K = B + T$, where $|T|_{X\to X} < \varepsilon$ and B is a finite-dimensional operator. Since A_n converges strongly to A, it follows that A_nB converges uniformly to AB. There exists a number $N(\varepsilon)$ such that

$$|A_nB - AB|_{X\to Y} < \varepsilon$$

for $n > N(\varepsilon)$. Thus, for any $\varepsilon > 0$,

$$|A_nK - AK|_{X\to Y} < |A_nB - AB|_{X\to Y} + |A_nT|_{X\to Y} + |AT|_{X\to Y} < C\,\varepsilon$$

$(n > N(\varepsilon))$. This implies the assertion of the lemma.

6.27. Now let $\hat{A}(z) \in \widehat{\mathfrak{A}_3^a/B(z)} \in \mathfrak{A}_3^a$. Before proceeding to approximation of the operator $\hat{A}(z)/B(z)$, we transform it into a form convenient for proof of uniform approxmation. We express the operators $R_b'(z)$, $b \subset a$, figuring in $\hat{A}(z)/B(z)$ as follows (see Eq. (4.36b)):

$$R_b'(z) = F_b'(z) \sum_{k=o}^{6} (-1)^k L_b'^k(z) + R_b'(z) L_b'^7(z) \tag{6.44}$$

We now express $F_b'(z)$ and $L_b'(z)$ in terms of $R_c'(z)$, $c \subset b$, then expressing the latter in the form (6.44), and so on. The final result is an expression of $\hat{A}(z)/B(z)$ as a linear combination of operators $\hat{A}^j(z)/B^j(z)$ in $\widehat{\mathfrak{A}_3^a}/\mathfrak{A}_3^a$, such that each operator $\hat{R}_b'(z)$ in the monomial $\hat{A}^j(z)$ is preceded by an operator $C_b(z)$ of $\mathfrak{A}_6^b$.

6.28. It follows from Proposition 5.1 and from our induction hypothesis (see the beginning of the proof of Proposition 6.3) that the operators $\hat{R}^b(z)$, $b \subset a$, satisfy the assumptions of Lemma 6.25, and so they have approximating sequences $\hat{R}_n^b(z)$, $n = 1,2,\ldots$, with the properties listed in Lemma 6.25. Set

$$\hat{R}_{b,n}(z)f = \hat{R}_n^b(z - \tau(p_b))f(p_b), \quad f \in L_2(R).$$

Let $v_{\alpha,n} \in S(R^3)$, $\psi_n^b \in S(R^b)$, be sequences such that

$$|v_{\alpha,n} - v_\alpha|_o \to 0 \quad \text{as} \quad n \to \infty, \tag{6.45}$$

$$|\psi_n^b - \psi^b|_b \to 0 \quad \text{as} \quad n \to \infty. \tag{6.46}$$

Let $V_{\alpha,n}$, $\pi_n(z)$ be the operators obtained from V_α, $\pi(z)$, respectively, when the functions v_α, ψ^b are replaced by $v_{\alpha,n}$, ψ_n^b. Now replace the operators V_α, $\pi(z)$ and $\hat{R}_b(z)$, $b \subset a$, in the formula for $\hat{A}(z)/B(z)$ by $V_{\alpha,n}$, $\pi_n(z)$ and $\hat{R}_{b,n}(z)$, respectively. Denote the operator thus defined by $\hat{A}_n(z)/B_n(z)$, $n = 1,2,\ldots$.

6.29. We shall prove that the operators $\hat{A}_n(z)$ and $B_n(z)$ are compact in $\hat{H}_{\sigma,\lambda}$ and $H_{\sigma,\lambda}$ respectively, and converge in the operator norm as $\operatorname{Im} z \to \pm 0$.

Lemma*). Let $Q(z)$ be the operator derived from a product of three monomials of type (6.15)--(6.17a) by omitting the resolvent on the extreme left and replacing the operators $R_b(z)$, $b \subset a$, and V_α by the operators $K_b(z)$ and W_α with the respective kernels

---

*) A similar proposition is proved in [4]. Our proof uses the same ideas as the proof of this proposition in [4].

$$\sum_{d\subseteq b} \frac{K_d^b(p^b,q^b;\ z - \tau(p_b))}{\lambda^d \pm \tau(p_d) - z}\,\delta(p_b - q_b),\quad K_d^b(p,q;\ z) \in C^\infty[\bar{C}^\pm; S(R^b \times R^b)],$$

and

$$w_\alpha(p^{u_\alpha} - q^{u_\alpha})\,\delta(p_{u_\alpha} - q_{u_\alpha}),\quad w_\alpha(k) \in S(R^3)\ .$$

Then the operator $Q(z)$, $z \in \Pi$, has a kernel satisfying the estimates

$$|Q(p,q;\ z)| < C(1 + |p - Gq|)^{-k} \sum \prod_{s=1}^{m} (1 + q_{i_s}^2)^{-1}, \tag{6.48}$$

$$|\Delta_z^t(u)\Delta_p^s(h)\Delta_q^r(\ell)Q(p,q;\ z)| \leqslant C\rho(p,q;\ \lambda)(1+|p- q|)^{-k} \sum \prod_{t=1}^{m} (1+q_{i_t}^2)^{-1}$$

$$\leqslant C\rho(p,q,\lambda)\ )\quad (1 + |p - Gq|)^{-k} \sum \prod_{t=1}^{m} (1 + q_{i_t}^2)^{-1}, \tag{6.49}$$

where $\det G > 0$, $\rho(\lambda) \in L_{\varepsilon^{-1}}(R^{6(m-1)})$, $\varepsilon = \theta(t+s + r - 1/2)(t+s + r - 1/2 + \delta)$, $\delta > 0$, $m = k(a)-2$ and the summation extends over all collections $i_1 \ldots i_m$ such that $i_k \neq i_s$ if $k \neq s$ for all $k > 0$.

PROOF. Expressing the kernel of $Q(z)$ in terms of the kernels of its constitutent operators, we obtain the following representation for $Q(p,q;\ z)$:*)

$$Q(p,q;\ z) = \int \frac{\Phi(x;\ z)\ dk}{\prod_{i=1}^{s} [\lambda_i + (x,\ R^i x) - z]},\quad x = (k,p,q), \tag{6.50}$$

---

*) This representation is quite obvious for operators of the form $W_{\alpha_i} R_o(z) W_{\alpha_2} \ldots R_o(z) W_{\alpha_k}$. Then the effect of an insertion of the operators K_b is easily traced and it only improves the estimates for the function ϕ.

where $\lambda_i \leq 0$, R^i, $i = 1,\ldots,s$, are symmetric nonnegative real matrices, $\sum_{i=1}^{s} R^i > 0$ and $\Phi(x;\, z)$ is an infinitely differentiable function whose derivatives satisfy the estimates

$$|D^\alpha \Phi(k,p,q;\, z)| < C(1 + |p - Gq|)^{-\beta}\, (1 + |k - Jq|)^{-\gamma} ,$$

for some rectangular matrix J and square matrix G. In Appendix III we shall consider matrices R^i satisfying conditions rather more general than those following directly from the form of $Q(z)$. There we shall study integrals of type (6.50) and establish estimates from which (6.48) and (6.49) will follow as special cases.

Here we shall outline the idea underlying the proof of these estimates. Using Feynman's method, we combine the s first-degree denominators in the integrand of (6.50) into a single s-th degree denominator and then, integrating s - 1 times by parts, lower the degree of this denominator to unity. The integral will then be studied by the usual methods of the theory of singular integrals.

The estimate thus obtained for the smoothness of the integral depends on the smoothness of the numerator and on the form of the denominator in the integral.

6.30. We now return to the operators $\hat{A}_n^j(z)$, $B_n^j(z)$. We have

$$\hat{A}_n^j(z) = \hat{R}_{b^j,n}(z) Q_n^j(z)\pi_n(z),\quad B_n^j(z) = \tilde{Q}_n^{j*}(z)\pi_n(z)\hat{R}_{c^j,n}(z),$$

where the operators $Q_n^j(z)$, $\tilde{Q}_n^j(z)$ satisfy the assumptions of Lemma 6.29.

Lemma 6.29 and the estimate

$$|\Delta_x^{\nu}(\ell)\Delta_y^{\sigma}(h)\phi|_{L_m(R^N)} \leq C \sup_u |\Delta_x^{\gamma\nu}(u)\phi|_{L_m(R^N)} + C \sup_v |\Delta^{\frac{\gamma\sigma}{\gamma-1}}(v)\phi|_{L_m(R^N)} ,$$

where $\gamma \geq 1$, $\phi(x,y) \in S(R^N)$, imply that $Q_n^j(z)\pi_n(z)$ and $\tilde{Q}_n^{j*}(z)\pi_n(z)$, $z \in \Pi$, are bounded operators from $\hat{H}_{\sigma,\eta,\lambda}$ to $H_{\sigma',\eta',\lambda}^{(b^j)}$ and $H_{\sigma',\eta',\lambda}$, $\sigma' > \sigma$, $\eta' > \eta$, respectively, continuous in $z \in \Pi$ in the uniform operator topology. This and the properties of the operators $\hat{R}_{b,n}(z)$, $b \subset a$, imply that

$$\hat{A}_n^j(z) \in \mathscr{L}(\hat{H}_{\sigma,\eta,\lambda}, \hat{H}_{\sigma',\eta',\lambda}) ,$$

$$B_n^j(z) \in \mathscr{L}(H_{\sigma,\eta,\lambda}, H_{\sigma',\eta',\lambda}), \ \sigma < \sigma', \ \eta < \eta', \quad z \in \Pi, \tag{6.51}$$

and are continuous in $z \in \Pi$ in the uniform operator topology.

Since the spaces $\hat{H}_{\sigma',\eta',\lambda}/H_{\sigma',\eta',\lambda}$ are compactly embedded in $H_{\sigma,\eta,\lambda}$, $\sigma < \sigma'$, $\eta < \eta'$, it follows from (6.51) that the operators $\hat{A}_n^j(z)/B_n^j(z)$ are compact in $\hat{H}_{\sigma,\eta,\lambda}/H_{\sigma,\eta,\lambda}$.

6.31. We can now prove that the operators $\hat{A}_n(z)/B_n(z)$ converge uniformly to $\hat{A}(z)/B(z)$. Recall that each operator $\hat{R}_b(z)$ appearing in the monomial $\hat{A}^j(z)/B^j(z)$ is preceded by an operator from $\mathfrak{A}_7^b$, which we have denoted by $C_b(z)$. Let

$$C_b(z) = D_b(z)E_b(z), \ D_b(z) \in \mathfrak{A}_6^b, \quad E_b(z) \in \mathfrak{A}_1^b(z), \tag{6.53}$$

and let $D^b(z)$ be the operator obtained from $D_b(z)$ when $R_c(z)$, V_α are replaced by $R_{c,b}(z)$, $c \subset b$, V_α^b (or, in other words, the restriction to $L_2(R^b)$): $D^b(z - \tau(p_b))f(p_b) = D_b(z)f$, $f \in L_2(R)$. Let $D_n^b(z)$ be the operator approximating $D^b(z)$ as a result of the previous substitution. Assume that $D_n^b(z)$ converges to $D^b(z)$ in $\mathscr{L}(H_{\sigma,\lambda}^b, H_{\sigma,\lambda}^b)$

Then $D^b(z)$ is also compact in $H^b_{\sigma,\lambda}$.

It follows from the relations (6.41)--(6.43) for the operators $\hat{R}^b_n(z)$, $n = 1,2,\ldots,\hat{R}^b(z)$, from Lemmas 6.10 for $a = b$, 6.25 and the compactness of $D^b(z)$, that

$$|(\hat{R}^b_n(z) - \hat{R}^b(z))D^b(z)|_{H^b_{\sigma,\lambda} \to \hat{H}^b_{\sigma,\lambda}} \to 0 \quad (n \to \infty), \tag{6.54}$$

$$\sup_{\lambda<\lambda^b-1} (1+|\lambda|)|D_\lambda(\hat{R}^b_n(z)D^b(z) - \hat{R}^b(z)D^b(z))|_{H_{\sigma,\eta}(R^b) \to \hat{H}^b_{\nu,\eta}} \to 0 \tag{6.55}$$

as $n \to \infty$. Similarly, we obtain

$$\int |\Delta^\sigma_\lambda(h)(\hat{R}^b_n(z)D^b(z) - \hat{R}^b(z)D^b(z)D^b(z))|^m_{H^b_{\sigma,\lambda} \to \hat{H}^b_\lambda} \frac{dh}{|h|} \to 0 \tag{6.56}$$

as $n \to \infty$.

Equations (6.47), (6.54)--(6.56) and Lemmas 6.10, 6.9 imply

$$|\hat{R}_{b,n}(z)C_b(z) - \hat{R}_b(z)C_b(z)|_{H^{(b)}_{\sigma,\lambda} \to \hat{H}^{(b)}_{\sigma,\lambda}} \to 0 \text{ as } n \to \infty . \tag{6.57}$$

It follows from Lemma 6.22 and Eqs. (6.45), (6.46), (6.57) that

$$|\hat{A}_n(z) - \hat{A}(z)|_{\hat{H}_{\sigma,\eta} \to \hat{H}_{\sigma,\lambda}} \to 0 \text{ as } n \to \infty,$$

$$|B_n(z) - B(z)|_{H_{\sigma,\lambda} \to H_{\sigma,\lambda}} \to 0 \text{ as } n \to \infty . \tag{6.58}$$

These relations, together with the observations that $B_n(z) = (V_{\alpha,n}R_o(z))^3$, where $u_\alpha = a \in A_{n-1}$, justifies (via induction) our assumption that the operators $D^b_n(z)$ converge uniformly to $D^b(z)$.

Since the set of compact operators in a Banach space is closed under uniform convergence, it now follows from (6.58), (6.59) and the compactness and continuity in z of the operators $\hat{A}_n(z)/B_n(z)$, $z \in \Pi$, that the operators $\hat{A}(z)/B(z)$, $z \in \Pi$, are compact and continuous in z, proving Lemma 6.24.

6.32. The operator $\tilde{F}(z)$ is investigated in exactly the same way as $L_b'(z)$. We have

LEMMA. The operator $\tilde{F}(z)$ may be expressed in the form

$$\tilde{F}(z) = \pi(z)\hat{F}(z),$$

where $\hat{F}(z)$ satisfies the estimates (6.11a)--(6.13a), (5.5a) for some δ.

6.33. Consider the equation

$$\hat{g} + \hat{L}(z)\,\hat{g} = \hat{F}(z)\phi, \quad \phi \in H_{\sigma,\lambda} \tag{6.60}$$

in the Banach space $\hat{H}_{\sigma,\lambda}$. Since $\hat{L}^3(z)$ is a compact operator in $\hat{H}_{\sigma,\lambda}$ the Fredholm alternative applies to this equation (see, e.g., [36], p. 579). Consider the corresponding homogeneous equation

$$\hat{g} + \hat{L}(z)\,\hat{g} = 0. \tag{6.61}$$

LEMMA. If Eq. (6.61) has a nontrivial solution $\hat{g}(\lambda \pm i0) \in \hat{H}_{\sigma,\lambda}$ for $z = \lambda \pm i0$ or $z = \lambda - i0$, $\lambda \neq \lambda^b$, $b \subset a$, $b \in A_{k(a)+1}$, $\sigma > 1/2$, then λ is an eigenvalue of the operator H, and $\pi(\lambda + i0)\hat{g}(\lambda + i0) =$ $(\pi(\lambda - i0)\hat{g}(\lambda - i0)$ is a corresponding eigenfunction.

PROOF. It is evident from Eq. (3.14) that it suffices to show that $F(z)^{-1}\pi(z)$, $z \in \Pi$: $\hat{H}_{\sigma,\lambda} \to H_{\sigma,\lambda}$ and the function $\pi(\lambda + i0)\hat{g}(\lambda + i0)$ $(\pi(\lambda - i0)\hat{g}(\lambda - i0)) \in D(H_o)$, i.e. it is square-integrable in a neighborhood of its singularities.

6.34. Using the same considerations as for the operators $L_b'(3)$ and $\hat{L}(z)$ and also the properties of $L_b(z)$ near the discrete spectrum of H^b

one obtains

LEMMA. The operator $F(z)^{-1}\pi(z)$, $z \in \Pi$, is bounded from $\hat{H}_{\sigma,\lambda}$ into $H_{\sigma,\lambda}$.

6.35. Now we proceed to the proof of integrability of $\pi(\lambda + i0)\hat{g}(\lambda + i0)/\pi(\lambda - i0)\hat{g}(\lambda - i0)$ near the singularities of $\pi(\lambda + i0)/\pi(\lambda - i0)$.

LEMMA. Let $\hat{g}(\lambda \pm i0) = \{g_b^{\pm}(p_b;\ \lambda),\ b \subset a,\ b \in \bar{A}\}$ be a solution of Eq. (6.61) for $z = \lambda + i0$ or $z = \lambda - i0$, in the space $\hat{H}_{\sigma,\lambda}$. Then

$$\int |g_b^{\pm}(p_b;\ \lambda)|^2\ \delta(\lambda^b + \tau(p_b) - \lambda)\ dp_b = 0,\ b \subset a,\ b \in \bar{A},\ \lambda \neq \lambda^b.$$

PROOF. To fix ideas, consider the case $z = \lambda + i0$. Define a vector

$$\hat{\phi}(\lambda,\varepsilon) = [\hat{L}(\lambda + i\varepsilon) - \hat{L}(\lambda + i0)]\hat{g}(\lambda + i0)$$

$$= \hat{L}(\lambda + i\varepsilon)\hat{g}(\lambda + i0) + \hat{g}(\lambda + i0)\ . \tag{6.62}$$

Since the operator $\hat{L}(z)$ is continuous with respect to $\operatorname{Im} z$ (see Lemma 6.22),

$$|\hat{\phi}(\lambda,\varepsilon)|_{\hat{H}_{\sigma,\lambda}} \to 0 \quad \text{as} \quad \varepsilon \to 0. \tag{6.63}$$

Applying the operator $\pi(\lambda + i\varepsilon)$ to both sides of (6.62), using (6.39) and introducing the notation $g(\lambda,\varepsilon) = \pi(\lambda + i\varepsilon)\hat{g}(\lambda + i\varepsilon)$, $\phi(\lambda,\varepsilon) = \pi(\lambda + i\varepsilon)\hat{\phi}(\lambda,\varepsilon)$, we obtain the equation

$$\phi(\lambda,\varepsilon) = L(\lambda + i\varepsilon)g(\lambda,\varepsilon) + g(\lambda,\varepsilon). \tag{6.64}$$

We now insert the formula (5.12) for $E + L(\lambda + i\varepsilon)$ into (6.64) and multiply the resulting equation by $F(\lambda + i\varepsilon)^{-1}$. This gives

$$(H - \lambda - i\varepsilon)g(\lambda,\varepsilon) = \phi_1(\lambda,\varepsilon), \tag{6.65}$$

where

$$\phi_1(\lambda,\varepsilon) = F(\lambda + i\varepsilon)^{-1}\phi(\lambda,\varepsilon) .$$

By virtue of (6.63) and Lemma 6.34 we have

$$|\phi_1(\lambda,\varepsilon)|_{H_{\sigma,\lambda}} \to 0 \quad \text{as} \quad \varepsilon \to 0. \tag{6.66}$$

Now consider the scalar product of (6.65) and $g(\lambda,\varepsilon)$. In view of the symmetry of H on $D(H_o)$, we obtain

$$-2i\varepsilon(g(\lambda,\varepsilon),\ g(\lambda,\varepsilon)) = \mathrm{Im}(\phi_1(\lambda,\varepsilon),\ g(\lambda,\varepsilon)). \tag{6.67}$$

The right-hand side of (6.67) is bounded in absolute value by $|\phi_1(\lambda,\varepsilon)|_{H_{\sigma,\lambda}}$, and therefore vanishes as $\varepsilon \to 0$. Writing out the left-hand side of (6.67), we have

$$-2i\varepsilon(g(\lambda,\varepsilon),g(\lambda,\varepsilon))$$

$$= - \sum_{b\subset a} \int 2i\varepsilon \frac{\psi^b(p^b)g_b^+(p_b;\ \lambda)}{\lambda^b + \tau(p_b) - \lambda - i\varepsilon} \frac{\overline{\psi}^c(p^c)\overline{g}_c^+(p_c;\ \lambda)}{\lambda^c + \tau(p_c) - \lambda + i\varepsilon} dp. \tag{6.68}$$

The terms with $b \neq c$ in (6.68) vanish as $\varepsilon \to 0$, since we may take the independent variables of integration to be, say, p_b and p^b; integration with respect to p^b eliminates the second singular denominator, and integration with respect to p_b does the same for the first; finally, in view of the continuity of singular integrals, we can let $\varepsilon \to 0$. Consider the diagonal terms in (6.68). We have

$$2i\varepsilon \sum \int \frac{\psi^b(p^b)g_b^+(p_b;\ \lambda)}{\lambda^b + \tau(p_b) - \lambda - i\varepsilon} dp^b\, dp_b$$

$$= 2i \sum \int |g_b^+(p_b;\ \lambda)|^2 \frac{\varepsilon}{(\lambda^b + \tau(p_b) - \lambda)^2 + \varepsilon^2} dp_b$$

$$\to 2i \sum \int |g_b^{+}(p_b;\ \lambda)|^2 \delta(\lambda^b + \tau(p_b) - \lambda)\ dp_b$$

as $\varepsilon \to 0$, and this implies the assertion of the lemma.

6.36. It is now quite easy to prove that the function $\pi(z)\hat{g}(z)$, $z = \lambda \pm i0$, $\sigma > 1/2$, where $\hat{g}(z)$ is a solution of Eq. (6.61), is square-integrable in the neighborhood of its singularities. Let

$$\pi(\lambda \pm i0)\hat{g}(\lambda \pm i0) = \sum \frac{\psi^b(p_b^b) g_b^{+}(p_b;\ \lambda)}{\lambda^b + \tau(p_b) - \lambda \mp i0} . \tag{6.69}$$

We shall show that

$$\int_{|\lambda^b+\tau(p_b)-\lambda|<\Delta} \frac{|\psi^b(p^b) g_b^{\pm}(p_b;\ \lambda)|^2}{|\lambda^b + \tau(p_b) - \lambda|^2}\ dp =$$

$$= \int_{|\lambda^b+\tau(p_b)-\lambda|<\Delta} \frac{|g_b^{\pm}(p_b;\ \lambda)|^2}{|\lambda^b + \tau(p_b) - \lambda|^2}\ dp_b < +\infty . \tag{6.70}$$

If $\lambda^b - \lambda > 0$, we have the estimate

$$\int_{|\lambda^b+\tau(p_b)-\lambda|<\Delta} \frac{|g_b^{\pm}(p_b;\ \lambda)|^2}{|\lambda^b + \tau(p_b) - \lambda|^2}\ dp_b < (\lambda^b - \lambda)^{-2} .$$

$$\int_{|\lambda^b+\tau(p_b)-\lambda|<\Delta} |g_b^{\pm}(p_b;\ \lambda)|^2 dp_b = C < \infty .$$

Let $\lambda = \lambda^b$, $b \notin A_{k(a)+1}$. Transform variables in (6.70) by $x = Ap_b$ so that the nonnegative quadratic form $\tau(p_b)$ is reduced to a sum of squares, and integrate with respect to the angles. The corresponding integral of

$g_b^{\pm}(p_b,\lambda)$ is bounded, since $g_b^{\pm} \in H_{\sigma,\lambda}(R_b)$. Hence we obtain

$$\int_{\tau(p_b)<\Delta} \frac{g_b^{\pm}(p_b;\ \lambda,f)|^2}{\tau^2(p_b)} dp_b \leq C \int_0^{\Delta} \frac{|x|^{3k-1} d|x|}{|x|^4} < \infty \quad ,$$

where $k = k(b) = k(a) \geq 2$ is the number of independent variables in p_b.

Finally, let $\lambda^b - \lambda < 0$. We take one of the integration variables to be $r = \tau(p_b)$, and the other a point ω on the surface Ω_b: $\tau(p_b) = 1$. A point $p_b \in R_b$ with spherical coordinates $r = \tau(p_b)$ and $\omega = p_b\ \tau(p_b)^{-1}$ will be written as $p_b = \sqrt{r}\omega$. Considering a function $f \in L_m(R_b)$ as a vector-function, $r \to f(\sqrt{r}\ \cdot)$, from R^+ into $L_m(\Omega_b)$ we define the family of operators

$$\Pi_b(r)\colon\ f \to f(\sqrt{r}\ \cdot) \quad , \tag{6.71}$$

acting from $L_m(R_b)$ into $L_m(\Omega_b)$. The result of Appendix IV implies that it is uniformly bounded from $H_\gamma(R_b)$ into $L_m(\Omega_b)$ and its σ-derivative, $\Delta_r^{\sigma}(h)\Pi(r)$, is uniformly bounded from $H_{\sigma,\gamma}(R_b)$ into $L_m(\Omega_b)$:

$$|\Delta_r^{\sigma}(h)\Pi_b(r)f|_{L_m(\Omega_b)} \leq C\, r^{-\frac{3}{2m}(k(b)-k(a))} |f|_{H_{\sigma,\gamma}(R_b)} \tag{6.72}$$

for any $b \in \bar{A}$ and $r > \gamma$.

Thus taking into consideration Lemma 6.35 we obtain

$$= \int_{|r-\lambda+\lambda^b|\leqslant\Delta} \frac{|(\Pi_b(r)g_b^{\pm}(\lambda,f))(\omega) - (\Pi_b(\lambda-\lambda^b)g_b^{\pm}(\lambda,f))(\omega)|^2}{|r-\lambda+\lambda^b|^2} r^{k-1}\, dr\, d\omega \lesssim$$

$$\lesssim C \int_{|r-\lambda+\lambda^b|\lesssim\Delta} \frac{|(\Pi_b(r) - \Pi_b(\lambda-\lambda^b))g_b^{\pm}(\lambda,\varepsilon)|^2_{L_m(\Omega_b)}}{|r-\lambda+\lambda^b|^2} r^{\frac{m}{2}(k-1)}\, dr\,, \tag{6.73}$$

where $k = \frac{3}{2}(k(b)-k(a))$. The last term in this inequality is finite in virtue of (6.72). This completes the proof of Lemma 6.33.

6.37. Now we formulate the second indirect restriction on H:

Condition B. The equations $\hat{f}+\hat{L}^a(\lambda^b \pm i0)\hat{f} = 0$, $b \in A_{k(a)+1}$, $b \subset a$, $a \in A'$, have no nontrivial solution in $\hat{H}^a_{\sigma,\lambda}$.

Using Lemmas 6.22, 6.24 and 6.33, Conditions A and B, and the relation

$$(\hat{E} + \hat{L}(z'))^{-1} - (\hat{E} + \hat{L}(z))^{-1} = -(\hat{E} + \hat{L}(z))^{-1}[\hat{L}(z') - \hat{L}(z)](\hat{E} + \hat{L}(z'))^{-1}, \tag{6.74}$$

we obtain: *)

LEMMA. For all points $z \in \Pi$ not in the discrete spectrum of H, the operator $\hat{E} + \hat{L}(z)$ has a bounded inverse in $\hat{H}_{\sigma,\lambda}$ which may be expressed as

$$(\hat{E} + \hat{L}(z))^{-1} = \hat{E} + \hat{B}(z), \tag{6.75}$$

where $\hat{B}(z)$ has the same properties (with $\delta = 0$) as $\hat{L}(z)$ in the $\Pi\setminus\{$a neighborhood of $\sigma_d(H)\}$.

It follows from Lemmas 6.37 and 6.32 that if z is in the region obtained from Π by deleting the points of the discrete spectrum of H together with certain neighborhoods thereof, then Eq. (6.60) has a solution $R(z)$ satisfying estimates (6.11a)--(6.13a). In conjunction with Proposition 5.1, this implies the assertion of Proposition 6.3, in order to complete the proof of Proposition 6.3 we should only prove Lemma 6.17.

6.38. Proof of Lemma 6.17. Estimate (6.35) means that $g(z)$ can be decomposed as follows

$$g(z) = \sum_{u_\alpha \subseteq e_{k+1}} g(z)^{\alpha}, \tag{6.76}$$

where $g(z)^{\alpha}$ satisfy the estimates of type (6.35) with the only α-term kept in the definition of $H_{\sigma,\lambda}^{(e_{k+1})}(R_v)$.

We will prove that there exists a pair $\bar{\alpha}$ satisfying $u_{\bar{\alpha}} \subseteq e_{k+1}$ such

*) We use here, that the spectrum of $\hat{L}(z)$ in $\hat{H}_{\sigma,\gamma,\lambda}$ does not depend on σ because $H_{\sigma',\lambda}$ is imbedded densely in $H_{\sigma,\lambda}$, $\sigma' > \sigma$ and $\hat{L}(z)$ is compact in $\hat{H}_{\sigma,\lambda}$.

that $g(z)$ obeys the estimate

$$|M_v^{e_{k+1}}(\eta_1) M_{e_{k+1}}(\eta)\, J^{-\nu}_{v,\bar{\alpha}}\, \chi_v(\gamma)\, g(z)|_{H_{\nu_o,\eta_o}(R^v)' \otimes L_m(R_v)} < C \tag{6.77}$$

and the similar estimates with $\Delta_z^\sigma(h)$ and $\Delta^\sigma(h)$, so decomposition (6.76) reduces to just one term. The induction hypothesis, (6.33 k+1), implies that $g(z)$ obeys all required estimates except of those for $v \not\subseteq e_k$, $v \subset e_{k+1}$.

Now we proceed to the proof of this estimate. The main part of this proof consists of the deduction of the handy representation for the function $g(z)$.

We consider monomial (6.29) with corresponding restrictions (6.30). Because of the condition $\bigcup_{k+1}^{t} u_{\alpha_i} \not\subseteq v$, it can be decomposed into a sum of monimials of the same type, say,

$$\overrightarrow{\prod_{i=n}^{r}} V_{\alpha_i} R_{c_i}(z) \tag{6.78}$$

satisfying the additional condition

$$\ell,\quad n < \ell < r \quad \bigcup_{i=n}^{\ell-1} (u_{\alpha_i} \cup c_i) \subseteq v, \quad u_{\alpha_\ell} \not\subseteq v \tag{6.79}$$

Indeed, any monomial of the type (6.78) which satisfies the condition

$$\bigcup_{i=n}^{t} u_{\alpha_1} \not\subseteq v \tag{6.80}$$

satisfies also one of the following two restrictions

$$p,\quad n \leqslant p \leqslant r \;:\; \bigcup_{i=n}^{p-1} (u_{\alpha_i} \cup c_i) \subseteq v,\quad u_{\alpha_p} \not\subseteq v\,,$$

$$\exists p,\quad n \leqslant p \leqslant r \quad : \quad \bigcup_{i=n}^{p-1} (u_{\alpha_i} \cup c_i) \cup u_{\alpha_p} \subseteq v, \qquad c_p \subseteq v$$

In the first case we obtain already the desired result. In the second case we represent the resolvent $R_{c_p}(z)$ in (6.78) as

$$R_{c_p}(z) = -L_{c_p}(z) R_{c_p}(z) + F_{c_p}(z)$$

and express $L_{c_p}(z)$ and $F_{c_p}(z)$ through the corresponding monomials. Thus monomial (6.78) in this case is decomposed of monomials of the same type. For each of these monomials we apply the same consideration as was done above. It is clear, since $\bigcup_{i=n}^{t} u_{\alpha_i} \nsubseteq v$, that on some step we obtain the desired decomposition of (6.78).

In order not to introduce new notations we assume that monomial (6.29) itself obeys the restriction

$$\exists \ell,\quad n \leqslant \ell \leqslant r: \qquad \bigcup_{i=n}^{\ell-1} (u_{\alpha_i} \cup c_i) \subseteq v, \quad u_{\alpha_\ell} \nsubseteq v \tag{6.81}$$

As was shown above this assumption does not affect the generality of the consideration.

It follows from (6.33 k+1) that

$$g(z) = \overset{\curvearrowright}{\prod_{i=k+1}^{t-1}} [V_{\alpha_i} R_{c_i}(z)] \phi \ ,$$

where

$$\phi = V_{\alpha_t} \pi(z) \{(R_{c_t}(z) f)_d, \quad d \subseteq c_t, \quad d \nsupseteq d_{k+1}\}$$

Using this representation and restriction (6.81), we obtain

$$P_v g(z) = (\psi^v, g(p_v; z)) =$$

$$= \Big(\overleftarrow{\prod_{i=k+1}^{\ell-1}} [R_{c_i,v}(\bar{z} - \tau(p_r)) V^v_{\alpha_i}] \psi^v, \ \Big(\overrightarrow{\prod_{i=\ell}^{t-1}} [V_{\alpha_i} R_{c_i}(z)] \phi\Big)(p_v)\Big)$$

Define furthermore

$$\bar{\phi}(z) = \overleftarrow{\prod_{i=k+1}^{\ell-1}} [R_{c_i,v}(z) V^v_{\alpha_i}] \psi^v ,$$

$$\chi(z) = \overrightarrow{\prod_{i=\ell}^{t-1}} [V_{\alpha_i} R_{c_i}(z)] \phi ,$$

$$P_J = \prod_{j_s \varepsilon J} P_{j_s} \quad \text{for} \quad J \subseteq (1,\dots,n)$$

Let $v = \{C_i\}$, k_i be the number of elements in the set C_i and

$$p^v = (p_i, \ i = 1\dots n), \quad p_v = (p_{C_i}, C_i \in v), \quad n_s : \ s \in C_{n_s} .$$

We obtain for any J satisfying $J \cap C_s = \{j_s\}$ $\forall C_s \in v$ and $i \in J$

$$P_v g(z) = \int \phi(p^v, z - \tau(p_v)) \chi(q) \prod_{s=1}^{n} \delta(q_s - p_s - k^{-1}_{n_s} p_{C_{n_s}}) \prod_{s=1}^{n} dq_s \, dp^v$$

$$= \int \phi(p^v, z - \tau(p_v)\chi(q)\delta(\sum_{s=1}^{n} q_s) \prod_{s \varepsilon J \setminus \{i\}} \delta(\sum_{t \varepsilon C_{n_s}} q_t - p_{C_{n_s}}) \times$$

$$\times \prod_{s \notin J} \delta(q_s - p_s - k^{-1}_{n_s} p_{C_{n_s}}) \prod_{s=1}^{n} dq_s \, dp^v$$

$$= \int (P_J \phi)(p^v, z - \tau(p_v))^J_{p_s \to q_s - k^{-1}_{m_s} p_{C_{n_s}}} (P_i \chi)(q^i) \prod_{s \varepsilon J \setminus \{i\}} \delta(\sum_{t \varepsilon C_{n_s}} q_t - p_{C_{n_s}}) \prod_{t \neq i} dq_t$$

(6.82)

The desired estimate (6.77) is implied now by (6.82) with J satisfying $J \cap d_\ell = d_\ell$ if one takes $\bar{\alpha} = \alpha_\ell$ and observes that in the domain

$$\lambda^v + \tau(p_v) > \gamma, \quad \gamma \in (\lambda - (1 - (4k(a))^{-1})\kappa, \lambda - (2k(a))^{-1}\kappa),$$

which is picked out by the factor $\theta_v(\gamma)$ in (6.77), $\lambda - \tau(p_v) < \mu^v$ and therefore (see Section 5) the function $\phi(z - \tau(p_v))$ satisfies the estimates

$$|D^k_{pv}\phi(z - \tau(p_v))|_v < C$$

(the definition of the norm $|\cdot|_v$ is given in Section 5). Lemma 6.15 is proved.

This completes the proof of Proposition 6.3.

6.39. Proposition 6.3 implies

Corollary. The operators $\hat{R}^a(z)$, $a \in A$, described in Proposition 6.3 satisfy the estimates

$$|\Delta^\alpha_z(h)R^a(z)|_{H_{s,\eta}(R^a)\to\hat{H}^a_{t,\theta}} < C(1 + |z|)^\delta ,$$

where $\hat{H}^a_{t,\theta} = \bigoplus_{b\subset a} H_{t,\theta}(R^a_b)$, the spaces $H_{s,\eta}(R^a_b)$ are defined in the end of Section 2, $s > t + \alpha$, $\alpha + t < \frac{1}{2}$, $\theta < \eta$.

The proof of this corollary follows from the estimates of Proposition 6.3 if one uses inequality (6.51) with $x = z$, $y = p$, $\nu = \alpha$, $\sigma = t$, $\gamma = \frac{\alpha+t}{\alpha}$, equipped with the corresponding weight functions and equality

$$M^a_b(p_b-h;\ \eta) \lesssim CM^a_b(p_b;\ \eta), \qquad |h| \lesssim 1,$$

and notes that

$$H_{s,\eta}(R^a) \subset H_{\sigma,\gamma,\theta}(R^a), \quad s > \sigma+\nu, \quad \eta > \theta - \frac{3}{m} .$$

7. PROOF OF THEOREMS 3.7 AND 3.8

In this section we shall again omit the index m of the eigenfunctions of H^a, $a \in \bar{A}$.

We begin with the proof of Eq. (3.10). Using (3.6), we obtain

$$W_{\pm} Hf = \{\lim_{\varepsilon\to\pm o} -i\varepsilon \int \overline{\psi}^a(p^a)(R(\lambda^a + \tau(p_a) + i\varepsilon)Hf)(p)dp^a,\ a \in \bar{A}\}$$

$$= \{(\lambda^a + \tau(p_a))(W_a^{(\pm)} f)(p_a),\ a \in \bar{A}\}$$

$$= \hat{H}W_{\pm} f. \tag{7.1}$$

We now prove that the operators $W_{\pm}$ are unitary. Let $f \in S(R)$ and

$$(R(z)f)(p,z) = \sum_{a\in\bar{A}} \frac{\psi^a(p^a)\phi_a(p_a,z,f)}{\lambda^a + \tau(p_a) - z} \tag{7.2}$$

It follows from the definition of $W_a^{(\pm)}$ that

$$(W_a^{(\pm)} f)(p_a) = \lim_{\varepsilon\to o} -i\varepsilon \sum_{b\in\bar{A}} \int \overline{\psi}^a(p^a) \frac{\psi^b(p^b)\phi_b(p_b,\lambda^a+\tau(p_a)\pm i\varepsilon,f)}{\lambda^b+\tau(p_b)-\lambda^a-\tau(p_a)\mp i\varepsilon} dp^a. \tag{7.3}$$

Let $b \neq a$. Then, since $\phi_b(p^b,z,f)$ satisfies a Hölder condition, the integral in (7.3) tends to a continuous function as $\varepsilon \to 0$, and therefore drops out thanks to the factor ε. If $b = a$, the limit of the integral, multiplied by $i\varepsilon$, is $\pm\ \phi_a(p_a,\lambda^a+\tau(p_a)\pm i0,\ f)$. Hence

$$(W_a^{(\pm)} f)(p_a) = \pm\ \phi_a(p_a,\lambda^a + \tau(p_a) \pm i0,\ f), \tag{7.4}$$

where $\phi_a(p_a,z,f)$ is defined by (7.2). It follows from the estimates (3.3) that

$\phi_a(p_a,\lambda^a + \tau(p_a) \pm i0,\ f) \in L_2(R_a^{\varepsilon})$, for all $\varepsilon > 0$, where

$$R_a^\varepsilon = \{p_a : p_a \in R_a, |\lambda^a + \tau(p_a) - \lambda^b| > \varepsilon, \quad b \in \bar{A}\}.$$

To prove that $W_\pm$ is an isometric operator from $L_2(R)$ to $\widehat{\mathfrak{A}}$, we use the relationship

$$\int_\sigma^\nu (f, dE_\lambda g) = \lim_{\varepsilon \to o} \int_\sigma^\nu (f, [R(\lambda + i\varepsilon) - R(\lambda - i\varepsilon)]g)d\lambda \,, \quad f,g \in L_2(R) \tag{7.5}$$

In view of (7.2), this yields

$$\int_\sigma^\nu (f, dE_\lambda g)$$

$$= \lim_{\varepsilon \to o} \sum_{a,b \in \bar{A}} 2i\varepsilon \int_\sigma^\nu \int \frac{\psi^a(p^a)\phi_a(p_a, \lambda \pm i\varepsilon, f)}{\lambda^a + \tau(p_a) - \lambda \mp i\varepsilon} \cdot \frac{\bar{\psi}^b(p^b)\bar{\phi}_b(p_b, \lambda \pm i\varepsilon, g)}{\lambda^b + \tau(p_b) - \lambda \pm i\varepsilon} \, dp \, d\lambda \,. \tag{7.6}$$

If $a \neq b$, the integral in (7.6) tends to a continuous function, since the variables of integration may be so chosen (say p^a, p_a) that integration over part of them (p^a) eliminates one singular denominator (the second), and integration over the other part removes the other singularity (the first). Thus terms in (7.6) with $a \neq b$ vanish thanks to the factor ε. Evaluating the contribution from the diagonal terms in (7.6), we get

$$\int_\sigma^\nu (f, dE_\lambda g) =$$

$$= \sum_{a \in \bar{A}} \int_\sigma^\nu \int \phi_a(p_a, \lambda \pm i0, f)\bar{\phi}_a(p_a, \lambda \pm i0, g)\delta(\lambda^a + \tau(p_a) - \lambda)dp_a d\lambda \,. \tag{7.7}$$

Substituting $\sigma = \mu_o = \inf H$ and $\nu = +\infty$ in (7.7), using the relations (7.4) and

$$\int_{\mu_o}^\infty dE_\lambda = E \,,$$

we obtain, for all ε,

$$\sum_{a\in\bar{A}\cap A'} \int_{R_a^\varepsilon} |\phi_a(p_a,\lambda^a + \tau(p_a) \pm i0, f)|^2 dp_a \leq (f,f)$$

and so

$$(f,f) = \sum_{a\in\bar{A}} (W_a^{(\pm)} f, W_a^{(\pm)} f). \tag{7.8}$$

Hence it follows that $W_\pm$ is isometric from $L_2(R)$ to $\widehat{\mathcal{A}}$.

We now prove that

$$W_\pm W_\pm^* = \hat{E}. \tag{7.9}$$

It will suffice to show that

$$W_\pm^a W_\pm^{b*} = \delta_{a,b} E_a, \quad a,b \in \bar{A} . \tag{7.10}$$

As we have shown above, the expression on the right of

$$W_\pm^a = \lim_{\varepsilon\to 0} i\varepsilon \int \delta(\lambda^a + T_a - \lambda) P_a R(\lambda \pm i\varepsilon) d\lambda \tag{7.11}$$

converges strongly as an operator from $L_2(R)$ into $L_2(R_a)$. Hence $W_\pm^{a*}$ is a weak limit of the adjoint expression:

$$W_\pm^{a*} f = \lim_{\varepsilon\to 0} - i\varepsilon \int R(\lambda \mp i\varepsilon)\delta(\lambda^a + T_a - \lambda) d\lambda \psi^a f ,$$

understood as an operator from $L_2(R_a)$ into $L_2(R)$. We want to prove that for any $f \in S(R_a)$, $g \in S(R_b)$

$$I = \varepsilon^2 \iint (R(\lambda \pm i\varepsilon)\delta(\lambda^a + T_a - \lambda)\psi^a f, R(\nu \mp i\varepsilon)\delta(\lambda^b + T_b - \nu)\psi^b g) d\lambda d\nu \to$$

$$\to \delta_{a,b}(f,g) \qquad (\varepsilon \to 0) \tag{7.12}$$

This implies the desired relation (7.10). During the proof of (7.12) we take for the definity one of the signs in (7.12), for example, + .

Using the Hilbert identity for resolvents we transform (7.12) as follows

$$I = -\varepsilon^2 \int (R_{T_b}(\lambda + 2i\varepsilon)R(\nu + i\varepsilon)\delta(\lambda^a + T_a - \lambda)\psi^a f, \psi^b g)d\lambda -$$

$$-\varepsilon^2 \int (\delta(\lambda^b + T_b - \nu)R(\nu - i\varepsilon)R_{T_a}(\nu - 2i\varepsilon)\psi^a f, \psi^b g)d\nu \tag{7.13}$$

The both terms on the right of (7.13) are considered in the same way, therefore we will demonstrate the further transformations on one of them - the second. Using the equality

$$R(z) = R_a(z) - R(z) \sum_{u_\alpha \not\subseteq a} V_\alpha R_a(z) \tag{7.14}$$

we obtain

$$R(\nu - i\varepsilon)R_{T_a}(\nu - 2i\varepsilon)P_a = (i\varepsilon)^{-1}(I - R(\nu - i\varepsilon) \sum_{u_\alpha \not\subseteq a} V_\alpha)$$

$$\times (R_{T_a}(\nu - i\varepsilon) - R_{T_a}(\nu - 2i\varepsilon))P_a$$

The function $V_\alpha R_{T_a}(z)\psi^a f$ can be reduced easily to an ordinary singular integral when $u_\alpha \not\subseteq a$. Applying standard theorems of the theory of one-dimension singular integrals we obtain that $V_\alpha R_{T_a}(z)\psi^a f \in H_{s,\eta}(R)$ and continuous (in $H_{s,\eta}(R)$) in $z \in C\setminus[\lambda^a,\infty)$. One can take here $s > 0$ and $\eta > \frac{3}{2}$. Hence

$$g_\varepsilon(\nu) = \sum_{u_\alpha \not\subseteq a} V_\alpha(R_{T_a}(\nu - i\varepsilon) - R_{T_a}(\nu - 2i\varepsilon))\psi^a f \in H_{s,\eta}(R)$$

and

$$|g_\varepsilon(\nu)|_{H_{s,\eta}(R)} \to 0 \qquad (\varepsilon \to 0) \tag{7.15}$$

Taking these into consideration we obtain

$$- \varepsilon \int (\delta(\lambda^b + T_b - \nu) R(\nu - i\varepsilon) R_{T_a}(\nu - 2i\varepsilon)\psi^a f, \psi^b g) d\nu =$$

$$= i\varepsilon \int [(\lambda^a + \tau(p_a) - \lambda^b - \tau(p_b) + i\varepsilon)^{-1} - (\lambda^a + \tau(p_a) - \lambda^b - \tau(p_b) + 2i\varepsilon)^{-1}]$$

$$\psi^a(p^a) f(p_a) \overline{\psi}^b(p^b) \overline{g}(p_b) dp - i\varepsilon \int (\delta(\lambda^b - T_b - \nu) R(\nu - i\varepsilon) g_\varepsilon(\nu), \psi^b g) d\nu$$

The second integral on the right side always vanishes because of (7.15) (here we use the information about $R(z)$ provided by Theorem 3.5). The first integral vanishes whenever $b \neq a$; if $b = a$ it gives the following contribution

$$\frac{1}{2} \int f(p_a) \overline{g}(p_a) dp_a = \frac{1}{2}(f,g) \tag{7.16}$$

The first expression in (7.13) provides the other half in (7.12). This completes the proof of Theorem 3.7.

We now proceed to the proof of Theorem 3.8. We have to show that $e^{iHt} e^{-i(\lambda^a + T_a)t} \psi^a \to W_\pm^{a*}$ as $t \to \pm\infty$ strongly in $L_2(R_a)$. Since $W_\pm^a$ and $e^{iHt} e^{-i(\lambda^a + T_a)t} \psi^a$ are isometric and $W_\pm^a$ intertwines H and $\lambda^a + T_a$, it is enough to prove that

$$e^{iT_a t} W_\pm^a e^{-iT_a t} \psi^a \to E_a \quad \text{as} \quad t \to \pm\infty \tag{7.17}$$

weakly in $L_2(R_a)$. We will prove the strong convergence in (7.17) directly, since a type of convergence in question does not effect our considerations.

Using (7.11) and (7.14) we obtain the following equation for $W_\pm^a$:

$$W_\pm^a = P_a + \sum_{u_\alpha \not\subset a} \int \delta(\lambda^a + T_a - \lambda) W_\pm^a V_\alpha R_a(\lambda \pm i0) d\lambda \tag{7.18}$$

It is not difficult to show that

$$\lim_{t \to \pm\infty} \int |\Delta_\lambda^\theta(h) V_\alpha R_{T_a}(\lambda \pm i0) e^{-iT_a t} \psi^a f|^2_{L_2(R)} d\lambda = 0, \quad u_\alpha \not\subset a, \tag{7.19}$$

for any $0 \leq \theta < 1$ and $f \in S(R)$. Here $\Delta_\lambda^o(h) = I$. Indeed, the function $V_\alpha R_{T_a}(z)\, e^{-itT_a} \psi^a f$ can be represented in form (II.7) (see Appendix II) with f replaced by $\exp(-iT_a t)\psi^a f$. This can be reduced after an appropriate substitution of the variable of integration to the integral

$$e^{i\lambda^a t} \int \frac{\chi_\alpha(s,p)}{s-z}\, e^{-its}\, ds ,$$

where χ_α satisfies the estimate

$$\int |\Delta_s^\gamma(h)\chi_\alpha(s)|^2_{L_2(R)} ds < C, \qquad 0 \leq \gamma < 1$$

In order to demonstrate that such integral obeys (7.19) we rewrite it as follows

$$\int \frac{\chi_\alpha(s,p)}{s-z}\, e^{-its}\, ds = 2\pi i\, e^{-itz} \int_{(t-\tau)\,\mathrm{Im} z \leq o} \hat{\chi}_\alpha(\tau,p) e^{itz} d\tau ,$$

where

$$\hat{\chi}_\alpha(\tau,p) = \int e^{-i\tau s}\, \chi_\alpha(s,p)$$

Taking for a definity $\mathrm{Im} z > 0$, i.e. $t > 0$, we obtain

$$\int |\Delta_\lambda^\theta(h) \int \frac{\chi_\alpha(s,p)}{s-\lambda-io}\, e^{-its} ds|^2\, d\lambda =$$

$$= 2\pi \int d\lambda | \int_t^\infty \frac{e^{ith}-1}{|h|^\theta}\, \hat{\chi}_\alpha(\tau,p) e^{it\lambda}\, d\tau|^2 +$$

$$+ 2\pi \frac{|e^{-ith}-1|^2}{|h|^{2\theta}} \int d\lambda | \int_t^\infty \hat{\chi}_\alpha(\tau,p) e^{it\lambda}\, d\tau|^2 \leq$$

$$\leq 2\pi \int_t^\infty \tau^{2\theta} |\hat{\chi}_\alpha(\tau,p)|^2 d\tau + 2\pi t^{2\theta} \int_t^\infty |\hat{\chi}_\alpha(\tau,p)|^2 d\tau \leq$$

$$\leq 2\pi\, t^{-2(\gamma-\theta)} \int_{-\infty}^{\infty} |\tau|^{2\gamma} |\hat{\chi}_\alpha(\tau,p)|^2\, d\tau$$

This gives

$$\int |\Delta_\lambda^\theta(h) \int \frac{\chi_\alpha(s,\cdot)}{s-\lambda-io}\, e^{-its} ds|^2_{L_2(R)}\, d\lambda \leq$$

$$\leq Ct^{-2(\gamma-\theta)} \sup_h \int |\Delta_s^\gamma(h)\chi_\alpha(s,\cdot)|^2_{L_2(R)}\, ds \,.$$

that proves (7.19).

We show now that (7.18) and (7.19) imply (7.17). Denote

$$g_\alpha(z,t) = W_\pm^a\, V_\alpha\, R_{T_a}(z)\, e^{-itT_a}\, \psi^a f$$

and introduce the function

$$G_\theta(\lambda) = \frac{1}{2\pi} \int e^{i\lambda\tau} (1+|t|^2)^{-\frac{\theta}{2}}\, d\tau$$

and the operator

$$(J_\lambda^\theta f)(\lambda) = \frac{1}{2\pi} \int e^{i\lambda\tau}\, \hat{f}(\tau)\, (1+|\tau|^2)^{\frac{\theta}{2}}\, d\tau \,,$$

$$\hat{f}(\tau) = \int e^{-i\lambda\tau}\, f(\lambda) d\lambda \,.$$

They satisfy

$$f(\lambda) = \int G_\theta(\lambda-\nu)\, (J_\nu^\theta f)(\nu) d\nu$$

Furthermore, $G_\theta(T_a-\gamma)$ denotes the operator of multiplication with the function $G_\theta(\tau(p_a)-\gamma)$. We have for $f \in S(R_a)$ and $\theta > \frac{1}{2}$

$$|\int \delta(\lambda^a+T_a-\lambda)\, g_\alpha(\lambda\pm i0) d\lambda| =$$

$$= |\int G_\theta(\lambda^a+T_a-\nu) J_\nu^\theta g_\alpha(\nu\pm i0,t) d\nu| \leq$$

$$\leq \{\int |G_{\theta}(\lambda)|^2 d\lambda\}^{\frac{1}{2}} \{\int |J_{\nu}^{\theta} g_{\alpha}(\nu \pm i0,t)|^2 d\nu\}^{\frac{1}{2}}$$

In order to complete the proof of (7.17) we note that

$$\int |G_{\theta}(\lambda)|^2 d\lambda < \infty, \quad \theta > \frac{1}{2},$$

$$J_{\lambda}^{\theta} g_{\alpha}(\lambda \pm i0,t) = W_{\pm}^{a} J_{\lambda}^{\theta} V_{\alpha} R_{T_a}(\lambda \pm i0) e^{-itT_a} \psi^a f$$

and

$$|J_{\lambda}^{\theta} f| \leq C(|f| + \sup_h |\Delta_{\lambda}^{\theta}(h) f|)$$

For the last inequality see book [34]. Theorem 3.8 is proved.

APPENDIX I

In this appendix we prove the estimate (5.7) of Section 5. In order to simplify somewhat notations we omit the superindex $\underline{a}$ from the formulas in the sequel. Let $f \in H_{\nu_o,\eta_o}(R^d) \otimes H_{\nu,\eta}(R_d)$, $\alpha = (i,j)$ and $e = d \cup u_\alpha$. The function $V_\alpha R_o(z)f$ is representable by the integral

$$V_\alpha R_o(z)f = \int \frac{v_\alpha(p_i-q)f(p)\Big|_{\substack{p_i \to q \\ p_j \to p_i+p_j-q}}}{\frac{1}{2m_i}q^2 + \frac{1}{2m_j}(p_i+p_j-q)^2 + \sum\limits_{k\neq i,j} \frac{1}{2m_k} p_k^2 - z} dq \tag{I.1}$$

Since the denominator of the integrand turns to zero nowhere, the smoothness estimates are trivial. We omit them here in order not to complicate the notations too much and concentrate completely on the decay estimates. Substituting for v_α and f their estimating function we obtain that (I.1) is majorized by the following integral

$$I = \int (q^2 + \sum_{k\neq i,j} p_k^2 + 1)^{-1} (1+|p_i-q|)^{-\eta_o} (N^d(\eta_o) N_d(\eta))(p)\Big|_{\substack{p_i \to q \\ p_j \to p_i+p_j-q}} dq \,. \tag{I.2}$$

The rest of this appendix is devoted to the estimation of this integral toward what we are proceeding now.

First we transform the estimating function to a suitable form such that the integral (I.2) then can be easily estimated. To this end we observe that using the inequality

$$(1+|a|)^{-\alpha}(1+|b|)^{-\alpha} < (1+|a-b|)^{-\alpha}[(1+|a|)^{-\alpha}+(1+|b|)^{-\alpha}] , \tag{I.3}$$

we can replace the estimating function $N(p_i, i \in I; \eta)$ by an equivalent function in which one previously specified variable, say p_k, appears in each summand at most in one factor:

$$N(p_j, j \in I; \eta)$$

$$\leq C[N^{(k)}(p_j, j \in I; \eta) + \sum_{\beta \subset I\setminus\{k\}} N_{(\beta)}(p_j, j \in I; \eta)(1+|p_k+\sum_{\ell\in\beta} p_\ell|)^{-\eta}] , \tag{I.4}$$

where

$$N^{(k)}(p_j, j \in I; \eta) = \sum_{\substack{\delta\in\Delta: \bigcup \delta_i \not\ni k \\ \delta_i\in\delta}} \prod_{\delta_i\in\delta} (1+|\sum_{\ell\in\delta_i} p_\ell|)^{-\eta} , \tag{I.5}$$

$$N_{(\beta)}(p_j, j \in I; \eta) = \sum_{\delta\in\Delta^{(\beta)}} \prod_{\delta_i\in\delta} (1+|\sum_{\ell\in\delta_i} p_\ell|)^{-\eta} , \tag{I.6}$$

$$\Delta^{(\beta)} = \{\delta \in \Delta_{I\setminus\{k\}} , \delta_j \cap \beta \in \{\phi, \delta_j\} , \delta_j \in \delta\} . \tag{I.7}$$

This may be done as follows. Suppose that the variable p_k occurs in three factors in each constituent summand of $N(p_j, j \in I; \eta)$. By the definition of the estimating function, these factors much have the form

$$(1 + |p_k + a|)^{-\eta}, \quad (1 + |p_k + a + b|)^{-\eta}, \quad (1 + |p_k + a + b + c|)^{-\eta} ,$$

where $b, c \neq 0$. Estimating their product, we have

$$(1 + |p + a|)^{-\eta}(1 + |p + a + b|)^{-\eta}(1 + |p + a + b + c|)^{-\eta}$$

$$< C(1 + |b|)^{-\eta}[1 + |p + a|)^{-\eta} + (1 + |p + a + b|)^{-\eta}](1 + |p + a + b + c|)^{-\eta}$$

$$< C(1 + |b|)^{-\eta}\{(1 + |b + c|)^{-\eta}[(1 + |p + a|)^{-\eta} + (1 + |p + a + b + c|)^{-\eta}]$$

$$+ (1 + |c|)^{-\eta}[(1 + |p + a + b|)^{-\eta} + (1 + |p + a + b + c|)^{-\eta}]\}$$

Products of more than three factors containing p_k are estimated similarly. On the other hand, it is obvious that estimation of $N(p_j, j \in I; \eta)$ in this way yields on the right (up to a scalar factor) a function $N(p_j, j \in I; \eta)$ none of whose summands contain the specified variable in more than one factor, which therefore is estimated by the function $N(p_j, j \in I; \eta)$.

Return now to the estimation of (I.2). If $u_\alpha \lesssim d$ then $e \equiv d \quad u_\alpha = d$ and the desired estimates of (I.2) could be derived easily. We consider here the case $u_\alpha \subseteq d$. Let $d = \{e_i\}$ and $i \in C_1$, $j \in C_2$.

We set $p_{C_2} = \sum_{C_s \in d, s \neq 2} p_{C_s}$ in the function $N_d(p_d; \eta)$ and estimate it according to (I.6) with the picked variable p_{C_1}. We obtain

$$N_d(p_d; \eta) < C\, N_e(p_e; \eta) \sum (1 + |p_{C_1} + \sum_{s \in \delta} p_{C_s}|)^{-\eta} \, . \,] \tag{I.8}$$

Using (I.8) and the Schwartz unequality we find

$$I \lesssim C\, N_e(p_e; \eta_1) \int \frac{(1+|p_i-q|)^{-\eta_1}}{1+q^2+\sum_{k\neq i,j} p_k^2} N^d(p; \eta) \Bigg|_{\substack{p_i \to q \\ p_j \to p_i + p_j - q}}$$

$$\times (1+|q+\sum_{s \in C_1, s \neq i} p_s + \sum_{s\ \delta} p_{C_s}|)^{-\eta}\, dq$$

$$\lesssim N_e(p_e;\ \eta)\Big[\int \frac{(1+|p_i-q)^{-2\eta_1}}{(1+q^2+\sum_{k\neq i,j} p_k^2)^2}\Big(N^d(p;\ \eta_1)\Big|_{\substack{p_i\to q\\ p_j\to p_i+p_j-q}}\Big)^2 dq\Big]^{1/2} \tag{I.9}$$

The integral in the square brackets of the right side of this expression is estimated as follows: We apply (I.4) with the chosen variable $p_k = p_1(=p_j)$ to the factor $N(p_\ell, \ell \in C_2;\ \eta)$ $(N(p_\ell, \ell \in C_2;\ \eta))$ in the function $N^d(p;\ \eta)$. This leads to the following estimates for $N^d(\eta)$:

$$\begin{aligned}
&N^d(p;\ \eta)\\
&< C \prod_3^{k(d)} N(p_\ell, \ell \in C_s;\ \eta)\{N^{(i)}(p_\ell, \ell \in C_1;\ \eta)N^{(j)}(p_\ell, \ell \in C_2;\ \eta)\\
&+ \sum_{\beta \subsetneq (k\in C_1, k\neq i)} N_{(\beta)}(p_\ell\ \ell \in C_1;\ \eta)[N^{(j)}(p_\ell, \ell \in C_2;\ \eta)\\
&+ \sum_{\gamma \subsetneq (k\in C_2, k\neq j)} N_{(\gamma)}(p_\ell, \ell \in C_2;\ \eta)(1+|p_i+p_j+\sum_{\ell\in\beta\cup\gamma} p_\ell|)^{-\eta}] \times\\
&\times (1+|p_i+\sum_{\ell\ \beta} p_\ell|)^{-\eta} + \sum_{\gamma \subsetneq (k\in C_2, k\neq j)} [N^{(i)}(p_\ell, \ell \in C_1;\ \eta)\\
&+ \sum_{\beta \subsetneq (k\in C_1, k\neq i)} N_{(\beta)}(p_\ell, \ell \in C_1;\ \eta) \times (1+|p_i+p_j+\sum_{\ell\in\beta\cup\gamma} p_\ell|)^{-\eta}]\\
&\times N_{(\gamma)}(p_\ell, \ell \in C_2;\ \eta)(1+|p_j+\sum_{\ell\in\gamma} p_\ell|)^{-\eta}\}\ ,
\end{aligned} \tag{I.10}$$

Now we substitute for $N^d(p;\ \eta)$ its estimate (I.10) three main terms in (I.10) give rise to the three integrals. Each of the obtained integrals is estimated separately. Since all of them can be estimated exactly in the same way, we consider here one of them, say, the integral generated by the second term in the right side of (I.10). Taking out from the integral the factors which are independent on k we conclude that it is necessary to estimate the following integrals

$$I_\beta = \int \frac{(1+|p_i-q|)^{-2n_1}(1+|q+\sum_{t\,\beta} p_t|)^{-2n_1}}{(1+q^2+\sum_{k\neq i,j} p_k^2)^2}\,dq$$

Using (I.5) and

$$\int (1+|q+a|)^{-\alpha}(1+|q|)^{-\beta}dq < C(1+|a|)^{-\min(\alpha,\beta)} ,$$

we obtain for I_β the estimate

$$I_\beta < C(1+|p_i + \sum_{s\,\beta} p_s|)^{-2\eta_1}$$

$$\times \int \frac{(1+|p_i-q|)^{-2\eta_1}+(1+|q+\sum_{t\,\beta} p_t|)^{-2\eta_1}}{(1+q^2+\sum_{k\neq i,j} p_k^2)^2}\,dq$$

$$\lesssim C(1+|p_i+\sum_{s\,\beta} p_s|)^{-2\eta_1}(1+\sum_{k\neq i,j} p_k^2)^{-2\eta_1}$$

These inequalities together with (I.10) give

$$I \lesssim CN^e(\eta_1)\,N_e(\eta)$$

what completes the estimation of (I.2).

APPENDIX II

This appendix presents the proof of Lemma 6.11 in Sec. 6. To begin with, we describe the properties of the spaces $H_{\sigma,\lambda}(R_b)$.

Let $\{\overset{\alpha}{X}\}$ be a collection of Banach spaces, all subspaces of some linear space; then $X = \sum \overset{\alpha}{X}$ is the Banach space whose elements are all possible sums $\sum \overset{\alpha}{x}$, $\overset{\alpha}{x} \in \overset{\alpha}{X}$, with norm $|x|_X = \inf_{x=\sum \overset{\alpha}{x}} \sum |\overset{\alpha}{x}|_{\overset{\alpha}{X}}$. It is clear that $X = \hat{X}/\hat{N}$, where

$$\hat{X} = \sum \oplus \overset{\alpha}{X}, \quad \hat{N} = \{\hat{x}\colon \ \hat{x} \in \hat{X}, \ \sum \overset{\alpha}{x} = 0\} .$$

Lemma I.1. Let $\{\overset{\alpha}{X}\}$ and $\{\overset{\alpha}{Y}\}$ be two collections of Banach spaces and A an operator with the property: for any α, there exists $\beta = \beta(\alpha)$ such that A is bounded from $\overset{\alpha}{X}$ to $\overset{\beta}{Y}$ (uniformly in α). Then A is bounded from $X = \sum \overset{\alpha}{X}$ to $Y = \sum \overset{\alpha}{Y}$.

Proof. Let $x \in X$, $x = \sum \overset{\alpha}{x}$, where $\overset{\alpha}{x} \in \overset{\alpha}{X}$, be an arbitrary decomposition of the element x. We have

$$|Ax|_Y = |A\sum \overset{\alpha}{x}|_Y \leq \sum |A\overset{\alpha}{x}|_Y \leq \sum |A\overset{\alpha}{x}|_{\overset{\beta(\alpha)}{Y}} \leq C\sum |\overset{\alpha}{x}|_{\overset{\alpha}{X}} .$$

Since C is independent of the decomposition $x = \sum \overset{\alpha}{x}$, this implies that

$$|Ax|_Y \leq C \inf_{x=\sum} \sum |\overset{\alpha}{x}|_{\overset{\alpha}{X}} = C|x|_X .$$

Q.E.D.

Using this lemma, we can rephrase the assertion of Lemma 6.11 slightly strengthening it as follows:

For any β the function

$$g_\alpha(z, \phi f) = V_\alpha R_{T_b}(z)\phi \tag{II.1}$$

satisfy the estimates

$$|g_\alpha(z, \phi f)|_{H^{\alpha(e)}_{\sigma,\lambda}} < C(1 + |z|)^\delta |\phi|_b \ |v_\alpha|_o \ |f|_{H^{\beta(d)}_{\sigma,\lambda}(R_b)} , \tag{II.2}$$

$$\int |\Delta^\sigma_\lambda(h) g_\alpha(z, \phi f)|^m_{H^{\alpha(e)}_\lambda} \frac{dh}{|h|} < C(1 + |z|)^{\delta m} |f|^m_{H^{\beta(d)}_{\sigma,\lambda}(R_b)} , \tag{II.3}$$

$$|g_\alpha(\lambda \pm i\varepsilon, \phi f) - g_\alpha(\lambda \pm i0, \phi f)|_{H^{\alpha(e)}_{\sigma,\lambda}}$$

$$+ \int |\Delta^\sigma_\lambda(h)(g_\alpha(\lambda \pm i\varepsilon, \phi f))|^m_{H^{\alpha(e)}_\lambda} \frac{dh}{|h|} \to 0 \tag{II.4}$$

as $\varepsilon \to 0$, for all $f \in H^{\beta(d)}_{\sigma,\lambda}$. Here $\lambda = \text{Re } z$ and $\phi \in H_{\nu_o,\eta_o}(R^b)$.

Inequality (II.3) is proved in the same way as (II.2), and in fact the appropriate estimates are simpler in form. The strong continuity with respect to $\text{Im } z$ follows automatically from the proof of (II.2). We shall therefore confine ourselves to proving the estimates (II.2).

We define a function $\chi(t) \in C^\infty\{[0,\infty)\}$ such that $\chi(t) = 0$ for $t > \text{Re } z - \lambda^b + 2$, $\chi(t) = 1$ for $t \lesssim \text{Re } z - \lambda^b + 1$, and set

$$f'(p_b) = f(p_b)\chi(\tau(p_b)), \ f''(p_b) = f(p_b)(1-\chi(\tau(p_b))) , \tag{II.5}$$

where $b' = b \cup u_\alpha$.

It follows from (II.5) that

$$f(p_b) = f'(p_b) + f''(p_b) \,. \tag{II.6}$$

It is an easy matter to estimate the function $g_{ij}(z, \phi f')$. This we now proceed to do. Let us expand (II.1) in detail. Let $b = \{C_1,\dots,C_s\}$, $i \in C_1$, $j \in C_2$, $M_{C_\ell} = \sum_{r \in C_\ell} m_r$, $\ell = 1,\dots,s$. We have

$$g_{ij}(p;z,\phi f) = \int \frac{v_{i,j}(p_{C_1}-k)\tilde{\phi}(p^b,p_{C_1},k)f(k,p_{C_2}+p_{C_2}-k,\, p_{C_3},\dots,p_{C_3})}{\lambda^b + \frac{1}{2M_{C_1}}k^2 + \frac{1}{2M_{C_2}}(p_{C_1}+p_{C_2}-k)^2 + \sum_{\ell=3}^{s}\frac{1}{2M_{C_\ell}}p_{C_\ell}^2 - z}\, dk \tag{II.7}$$

where $\tilde{\phi}(p^b,p_{C_1},k) = \phi(p^b)\Big|_{\substack{p_i \to p_i - p_{C_1} + k \\ p_j \to p_j + p_{C_1} - k}}$. It follows that the denominator in (II.7) is not singular in the region where $f''(p)$ does not vanish, i.e. for

$$\lambda^b + \tau(p_{b'}) - \lambda \geqslant 1$$

Thus the function $g_{ij}(p;\, z,\, \phi f'')$ has the same order of smoothness as the numerator of the integrand. Consequently, in order to derive the necessary estimates we need only to estimate the behavior of the integral for large values of the external variables p.

The integral (II.7) and its Hölder differences are majorized by the expression

$$I = \int \frac{(1+|p_{C_1}-k|)^{-\eta_1} N^b(p;\eta_1)\Big|_{\substack{p_i \to p_i - p_{C_1}+k \\ p_j \to p_j + p_{C_1} - k}}}{1 + k^2 + (p_{C_1}+p_{C_2})^2 + \sum_{i=3}^{s} p_{C_i}^2} N_b^d(p_b;\eta_1) \times N_d(p_d;\eta)\Big|_{\substack{p_{C_1} \to k \\ p_{C_2} \to p_{C_1} + p_{C_2} - k}} dk \tag{II.8}$$

Now we estimate this integral. It follows from the definition of the estimating functions that

$$N_c^b N_b^d \leq N_c^d, \qquad c \subset b \subset d$$

Using this inequality we find

$$N^b(p;\,\eta)\Big|_{\substack{p_i \to p_i - p_{C_1} + k \\ p_j \to p_j + p_{C_1} - k}} \quad N_b^d(p_b;\,\eta)\Big|_{\substack{p_{C_1} \to k \\ p_{C_2} \to p_{C_1} - k}} \tag{II.9}$$

$$= (N^b(p;\,\eta)\, N_b^d(p_b;\,\eta))\Big|_{\substack{p_i \to p_i - p_{C_1} + k \\ p_j \to p_j + p_{C_1} - k}} < N^d(p;\,\eta)\Big|_{\substack{p_i \to p_i - p_{C_1} + k \\ p_j \to p_j + p_{C_1} - k}}$$

Thus integral (II.8) is majorized as follows:

$$I \leq \int (k^2 + \sum_{s \neq 1,2} p_{C_s}^2 + 1)^{-1} (1 + |p_{C_1} - k|)^{-\eta_0} (N^d(\eta_1) N_d(\eta))\Big|_{\substack{p_i \to p_i - p_{C_1} + k \\ p_j \to p_j + p_{C_1} - k}} dk \tag{II.10}$$

The integral in the right-hand side of this inequality differs from integral (I.2), studied in Appendix I, essentially only by the first factor in the integrand. This can be easily checked substituting in (II.10) the variable of the integration as follows:

$$k = q + \sum_{s \in C_1 \setminus \{i\}} p_s$$

The same estimate, as for (I.2), is valid also for (II.10)

$$I \leqslant CN^e(\eta_1)N_e(\eta)$$

This completes the proof of the desired estimates for $g_\alpha(z,\phi f")$.

We now estimate $g_\alpha(z,\phi f')$ with f' given in (II.5). To avoid cumbersome expressions, we shall confine attention to the case $b \in A_n$, i.e. $b = \{(1)\ldots(n)\}$. The other cases are considered in wholly analogous fashion. To simplify the notation, we shall also assume that $a \in A_1$.

Thus, consider the function

$$g_{ij}(z,f') = V_{ij}\ R_0(z)\ f'. \tag{II.11}$$

This function has compact support with respect to all variables except $p_i - p_j$; the dependence of the function on the latter is determined only by v_{ij}. It will therefore suffice to prove smoothness estimates for $g_{ij}(z,f')$.

For this reason, throughout the estimation procedure we shall ignore the estimating functions, i.e., we shall set $\eta = 0$. Then the spaces become

$$\overset{\alpha}{H} = \overset{\alpha}{H}{}^m_\nu = J^\nu_\alpha L_m(R) \tag{II.12}$$

where, remind,

$$J^\nu_\alpha f = \frac{1}{(2\pi)^{3(n-1)}} \int e^{ipx}\,\hat{f}(x)\,(1+|x_i-x_j|^2)^{-\frac{\nu}{2}}\,dx\ ; \qquad \alpha = (i,j)\ ,$$

$$\hat{f}(x) = \int e^{-ipx}\ f(p)dp\ . \tag{II.13}$$

and

$$\overset{\alpha}{H}_\sigma = \overset{\alpha}{H}{}^m_{\sigma,\nu} = \{f \in \overset{\alpha}{H}{}^m_\nu\ ;\ \int |\Delta^\sigma(h)f|^m_{\overset{\alpha}{H}{}^m_\nu} \frac{dh}{|h|^{3(n-1)}} \leqslant C\}\ ,\quad \sigma > 0 \tag{II.14}$$

We mention a property of the spaces $\overset{ij}{H}_{\sigma,\lambda}$ which follow directly from their definitions.

$$\overset{ij}{H}{}^{m}_{\sigma',\nu'} \subset \overset{ij}{H}{}^{m}_{\sigma,\nu} \;, \quad \sigma' \geqslant \sigma, \quad \nu' \geqslant \nu \;. \tag{II.15}$$

We have to prove that the function

$$g_\alpha(z,f) = V_\alpha R_0(z) f \tag{II.16}$$

satisfies the estimates

$$|g_\alpha(z,f)|_{\overset{\alpha}{H}} \leqslant C|v_\alpha|_\tau (1 + |z|)^\delta |f|_{\overset{\beta}{H}} \;, \tag{II.17}$$

$$|g_\alpha(z,f)|_{\overset{\alpha}{H}_\sigma} \leqslant C|v_\alpha|_{\tau+\sigma} (1 + |z|)^\delta |f|_{\overset{\beta}{H}_\sigma} \;, \tag{II.18}$$

where $\nu > 3/m$, $\tau > \nu + m^{-1}$, $\sigma < 1/2 + 1/m$ and

$$|v|_\tau = |(1+|\cdot|)^{1/2} J^{-\tau} v|_{L_m(R^3)}$$

Throughout our proof of these estimates, we shall limit ourselves to the case of unit masses:

$$m_1 = \dots = m_n = 1.$$

We put

$$\overset{i_1 \dots i_s}{p} = (p_k,\ 1 \leqslant k \leqslant n,\ k \neq i_1 \dots i_s),\ (\overset{i}{P_i} f)(p) = f(p)\Big|_{p_1 = -\sum_{k\neq i} p_k}$$

and

$$\overset{ij}{\tau(p)} = \tau(p_{u_{ij}})\Big|_{p_i + p_j = -\sum_{k \neq i,j} p_k}$$

Investigation of the function (II.16) gives separate consideration to the various possible relationships between α and β. We assume throughout that $\alpha = (i,j)$.

a) $\alpha = \beta$. Recalling the definition of the operators V_α and $R_0(z)$, we obtain the following expression for the function (II.16):

$$(P_i\, g_\alpha(z,f))(\overset{i}{p};\, z)$$

$$= \int \frac{v_\alpha(p_j - q_j)\,\delta(p_i + p_j - q_i - q_j) \prod_{k\neq i,j} \delta(p_k - q_k)}{1/2q_1^2 + \ldots + 1/2q_n^2 - z} (P_i f)(\overset{i}{q})\, d^n q$$

$$= \int \frac{v_\alpha(p_j - q_j)(P_i f)(q_j, \overset{ij}{p})\, dq_j}{1/2(q_j + \sum_{k\neq i,j} p_k)^2 + 1/2q_j^2 + 1/2 \sum_{k\neq i,j} p_k^2 - z} \tag{II.19}$$

We make the substitution $k = q_j = 1/2 \sum_{s\neq i,j} p_s$ in the integral (II.19). This gives

$$P_i g_\alpha(z,f) = \int \frac{v_\alpha(p_j + 1/2 \sum_{s\neq i,j} p_s - k)}{k^2 + \tau(\overset{ij}{p}) - z} (P_i f)(k - 1/2 \sum_{s\neq i,j} p_s,\, \overset{ij}{p})\, dk. \tag{II.20}$$

Set

$$\phi(k, \overset{i}{p}) = v_\alpha(p_j + 1/2 \sum_{s\neq i,j} p_s - k)(P_i f)(k - 1/2 \sum_{s\neq i,j} p_s,\, \overset{ij}{p}).$$

If $f \in \overset{\alpha}{H}$, the function $\phi(k, \overset{i}{p})$ satisfies the estimate

$$\int |\Delta_k^\gamma(h)\phi(k)|^m_{\overset{\alpha}{H}}\, dk \leq C|v_\alpha|^m_{\nu_1}\, |f|^m_{\overset{\alpha}{H}}, \quad 0 \leq \gamma \leq \min(\nu, \nu_1 - \nu) \tag{II.21}$$

Consider the function

$$\int \phi(\sqrt{t}\omega, \overset{i}{p})\, d\omega, \quad t \geq 0. \tag{II.22}$$

We extend this function to the region $(-\infty, 0)$ of the t-axis as a smooth function decreasing rapidly at infinity. Denote the resulting function by $\tilde{\phi}(t, \overset{i}{p})$. In terms of this function, the integral (II.20) becomes

$$P_i\, g_\alpha(z,f) = \int \frac{\tilde{\phi}(t,\overset{i}{p})\, t_+^{1/2}}{t + \tau(\overset{ij}{p}) - z}\, dt. \tag{II.23}$$

We now establish estimates for the function $\tilde{\phi}(t, \overset{i}{p})$, $t \geq 0$.

Lemma II.2. Let $f(k) \in S(R^3)$. Then

$$\int\int_0^\infty \left| \int f(\sqrt{t + h}\,\omega)\,d\omega - \int f(\sqrt{t}\,\omega)\,d\omega \right|^m t^{1/2(sm+1)}\, dt\, \frac{dh}{|h|^{1+sm}}$$

$$\leq C \int\int |f(p + h) - f(p)|^m\, dp\, \frac{dh}{|h|^{3+sm}}\,. \tag{II.24}$$

Proof. Let $f \in S(R^3)$. Then the integral on the left of (II.24) exists, and we may change the order of integration with respect to t and h. Doing this and then making the substitution $h \to \ell = \sqrt{t + h} - \sqrt{t}$, we get

$$\int_0^\infty\int_0^\infty \left| \int [f(\sqrt{t + h}\,\omega) - f(\sqrt{t}\,\omega)]\,d\omega \right|^m t^{1/2(sm+1)}\, dt\, \frac{dh}{h^{1+sm}}$$

$$\leq \int_0^\infty\int_0^\infty\int |f(\sqrt{t + h}\,\omega) - f(\sqrt{t}\,\omega)|^m\, d\omega\, \frac{dh}{h^{1+sm}}\, t^{1/2(sm+1)}\, dt$$

$$= \int_0^\infty\int_0^\infty\int |f(\sqrt{t}\,\omega + \ell\omega) = f(\sqrt{t}\,\omega)|^m\, d\omega\, \frac{2(\ell + \sqrt{t})}{(\ell + 2\sqrt{t})^{1+sm}}\, \frac{d\ell}{\ell^{1+sm}}\, t^{1/2(sm+1)}\, dt$$

$$\leq \int_0^\infty\int |f(p + \ell \tfrac{p}{|p|}) - f(p)|^m\, dp\, \frac{d\ell}{\ell^{1+sm}}$$

$$\leq \int_0^\infty \int |(p + \ell\frac{p}{|p|}) - f(p + \ell\frac{p'}{|p|})|^m \, dp \, \frac{d\ell}{\ell^{1+sm}} +$$

$$+ \int_0^\infty \int |f(p + \ell\frac{p'}{|p|}) - f(p + \ell\frac{p''}{|p|})|^m \, dp \, \frac{d\ell}{\ell^{1+sm}} +$$

$$+ \int_0^\infty \int |f(p + \ell\frac{p''}{|p|}) - f(p)|^m \, dp \, \frac{d\ell}{\ell^{1+sm}} , \tag{II.25}$$

where $p' = (p_1, p_2, 0)$, $p'' = (p_1, 0, 0)$, $(p_1, p_2, p_3) = p$.

We first estimate the first integral on the right of the chain of inequalities (II.25). We apply the substitution $q_1 = p_1 + \ell p_1/|p|$. We have

$$dq_1 = dp_1 (1 + \ell\frac{p_2^2 + p_3^2}{|p|^3}) . \tag{II.26}$$

Let α_i denote the function $p_i/|p|$, $i = 1,2,3$, expressed in terms of the variables q_1, p_2, p_3. Omitting the second term in parentheses in (II.26), we obtain

$$\int_0^\infty \int |f(p + \ell\frac{p}{|p|}) = f(p + \ell\frac{p'}{|p|})|^m \, dp \, \frac{d\ell}{\ell^{1+sm}}$$

$$\leq \int_0^\infty \int |f(q_1, p_2+\alpha_2\ell, p_3+\alpha_3\ell) - f(q_1, p_2+\ell\alpha_2, p_3)|^m \, dq_1 dp_2 dp_3 \, \frac{d\ell}{\ell^{1+sm}} . \tag{II.27}$$

We now transform the next variable by $p_2 \to q_2 = p_2 + \ell\alpha_2$. Then

$$dq_2 = dp_2 \left\{ 1 + \frac{p_1^2 + p_3^2 - \ell\frac{p_1^2 p_1^2}{|p|^3} (1 + \frac{p_2^2 + p_3^2}{|p|^3})^{-1}}{|p|^3} \right\} , \tag{II.28}$$

where p_1 is to be expressed in terms of q_1, p_2, p_3. Note that the second term in the bracketed expression in (II.28) is nonnegative. Therefore, in estimating the integral on the right of inequality (II.27) we may replace the bracketed term in (II.28) by unity. This gives

$$\int_0^\infty \int |f(q_1,p_2+\ell\alpha_2,p_3+\ell\alpha_3) - f(q_1,p_2+\ell\alpha_2,p_3)|^m dq_1 dp_2 dp_3 \frac{d\ell}{\ell^{1+sm}}$$

$$\leq \int_0^\infty \int |f(q_1,q_2,p_3+\ell\beta) - f(q_1,q_2,p_3)|^m dq_1 dq_2 dp_3 \frac{d\ell}{\ell^{1+sm}}$$

$$\leq \int_{-\infty}^\infty \int |f(p+ue_3) - f(p)|^m dp \frac{du}{|u|^{1+sm}} . \tag{II.29}$$

Here β denotes the function α_3 expressed in terms of q_1,q_2,p_3.

To estimate the second integral on the right of (II.25), we make one substitution $p_2 \to q_2 + p_2/|p|$ and then proceed in the same way as for the second integral in (II.29). Estimation of the third integral on the right of (II.25) involves only the last operation. This completes the proof of the lemma.

Lemma II.2 implies an estimate for the function (II.22):

$$\int_0^\infty \int_0^\infty |\Delta_t^\delta(h)\tilde{\phi}(t)|_{H}^{m}{}_{\alpha} \, t^{1/2(\delta m+1)} \, dt \frac{dh}{h} < C|v_\alpha|_\tau^m |f|_{H}^{m}{}_{\alpha} , \quad 0 \leq \delta \leq \min(\nu,\tau-\nu). \tag{II.30}$$

We now estimate the integral (II.23). We have

$$\Delta_\lambda^\delta(h) P_i g_a(z,f)$$

$$= \int_{-\infty}^\infty \frac{\Delta_t^\delta(h)\tilde{\phi}^i(t,p)}{t+\tau^{ij}(p)-z} t_+^{1/2} dt + |h|^{-\delta} \int_{-\infty}^\infty \frac{\tilde{\phi}^i(t+h,p)[(t+h)_+^{1/2} - t_+^{1/2}]}{t+\tau^{ij}(p)-z} dt$$

(II.31)

Using (II.31), the inequality

$$|(t+h)_+^{1/2} - t_+^{1/2} - t_+^{1/2} < C|h|^s|t|^{1/2-s}, \quad 1/2 \leq s \leq 1, \tag{II.32}$$

theorems on singular integrals in L_m-spaces (see, e.g., [37], [34]), and elementary embedding theorems for Sobolev and Hölder spaces (see inequality (S.3)), we obtain the estimate

$$\begin{aligned} &|J_{p_j}^{-\nu}\, \Delta_\lambda^\theta(\ell) P_i g_\alpha(z,f)| \\ &\leq \sup_u \sum_{k=o}^{1} \{\int |J_{p_j}^{-\nu}\, \Delta_\lambda^{\delta k}(u)\; P_i g_\alpha(z,f)|^m\, d\lambda\}^{1/m} \\ &\leq \sup_u \sum_{k=o}^{1} \{\int |J_{p_j}^{-\nu}\, \Delta_t^{\delta k}(u)\tilde{\phi}(t,\overset{i}{p})|^m\, |t|^{1/2mk}(1+|t|^{1/2m(1-k)})dt\}^{1/m} \end{aligned} \tag{II.33}$$

where $\delta > \frac{1}{m} + \theta$. Integrating the m-th power with respect to $\overset{i}{p}$, and using estimate (II.30), we obtain

$$|\Delta_\lambda^\theta(h)\; g_\alpha(z,f)|_{H^\alpha} \leq C|v_\alpha|_\tau\, |f|_{H^\alpha}, \qquad 0 \leq \theta < \tau-\nu-\frac{1}{m} > 0. \tag{II.34}$$

whence follows the estimate (II.17).

We now prove the estimate (II.18) for $\alpha = \beta$. Let $f \in H_\sigma^\alpha$. Lemma II.2 yields an estimate for the function $\tilde{\phi}(t,\overset{i}{p})$:

$$\int_o^\infty\int_o^\infty |\Delta_t^\delta(h)\tilde{\phi}(t)|_{H_\sigma^\alpha}^m\; t^{1/2(\delta m+1)}(1+t)^{\frac{m}{2}(1-\delta)}\, dt\,\frac{dh}{h} \leq C|v_\alpha|_{\tau+\sigma}^m |f|_{H_\sigma^\alpha}^m \tag{II.35}$$

Next, we have

$$\Delta_{p_j}^\sigma(h)P_i g(z,f) = \int \frac{\Delta_{p_j}^\sigma(h)\tilde{\phi}(t,\overset{i}{p})}{t+\tau(\overset{ij}{p})-z}\, t_+^{1/2}\, dt\,. \tag{II.36}$$

Hence, using (II.34), we obtain

$$|\Delta^{\theta}_{\lambda}(\ell)\Delta^{\sigma}_{p_j}(h)g_{\alpha}(z,f)|_{\substack{\alpha\\ H}} \leq C|v_{\alpha}|_{\tau+\sigma}\, |\Delta^{\sigma}_{p_j}(h)f|_{\substack{\alpha\\ H}} \,. \tag{II.37}$$

Now let $k \neq i,j$. We have $(\overset{ij}{p} = \sum p_k e_k)$

$$\Delta^{\sigma}_{p_k}(h)P_i g_{\alpha}(z,f) = \int_{-\infty}^{\infty} \frac{\Delta^{\sigma}_{p_k}(h)\tilde{\phi}(t,\overset{i}{p})t_+^{1/2}}{t + \tau(\overset{ij}{p} + he_k) - z}\,dt \,+$$

$$+\,|h|^{-\sigma}\Big\{\int_{-\infty}^{\infty} \frac{\tilde{\phi}(t,\overset{i}{p})t_+^{1/2}\,dt}{t + \tau(\overset{ij}{p} + he_k) - z} - \int_{-\infty}^{\infty} \frac{\tilde{\phi}(t,\overset{i}{p})t_+^{1/2}\,dt}{t + \tau(\overset{ij}{p}) - z}\Big\} = g' + g'' \,. \tag{II.38}$$

Applying inequality (II.34) to the first term on the right of (II.38), we obtain the estimate

$$|\Delta^{\theta}_{\lambda}(\ell)g'(z)|_{\substack{\alpha\\ H}} < C|v_{\alpha}|_{\tau+\sigma}\, |\Delta^{\sigma}_{p_k}(h)f|_{\substack{\alpha\\ H}} \,, \qquad 0 \leq \theta < \tau-\nu-\frac{1}{m} \,. \tag{II.39}$$

To estimate the second term on the right of (II.38), we consider two cases:

I. $\tau(\overset{ij}{p}) > 10(1+\lambda)$. In this case the integrals in braces in (II.38) are nonsingular for $|h| < 1$, and we use the inequality

$$|(t + \tau(\overset{ij}{p} + he_k) - \lambda)^{-1} - (t + \tau(\overset{ij}{p}) - \lambda)^{-1}| < C|h|(t + \tau(\overset{ij}{p}) - \lambda)^{-1} \,.$$

This gives the estimate

$$|D_z\, g''(z)|_{\substack{\alpha\\ H}} \leq C|v_{\alpha}|_{\nu,\eta}|f|_{\substack{\alpha\\ H}} \,. \tag{II.40}$$

II. $\tau(\overset{ij}{p}) < 10(1+\lambda)$. Denote $u = \tau(\overset{ij}{p} + he_k) - \tau(\overset{ij}{p})$. Making the substitution $t \to \tau = t + u$ in the first integral in braces in (II.38), we bring the second term on the right of (II.38) to the form

$$\overset{i}{g}''(p;\ z) = |h|^{-\sigma} \int \frac{\Delta_t(-u)\ \overset{i}{\tilde{\phi}}(t,p)}{t + \underset{ij}{\tau}(p) - z}\ t_+^{1/2}\ dt +$$

$$+ \int_{-\infty}^{\infty} \frac{\overset{i}{\tilde{\phi}}(t-u,p)}{t + \underset{ij}{\tau}(p) - z} \times \frac{(t-u)_+^{1/2} - t_+^{1/2}}{|h|^{\sigma}}\ dt \qquad \text{(II.41)}$$

Let us estimate the first integral in this expression. It follows from embedding theorems for spaces of differentiable functions (see Supplement, formula (S.3)) and theorems for singular integrals in L_m (see formula (S.5)) that

$$\left|\int_{-\infty}^{\infty} \frac{\Delta_t(-u)\ \overset{i}{\tilde{\phi}}(t,p)}{t + \underset{ij}{\tau}(p) - z}\ t_+^{1/2}\ dt\right|$$

$$\lesssim \sup_{\ell} \sum_{k=0}^{1} \left\{\int \left|\Delta_\lambda^{\delta k}(\ell) \int_{-\infty}^{\infty} \frac{\Delta_t(-u)\ \overset{i}{\tilde{\phi}}(t,p)}{t + \underset{ij}{\tau}(p) - z}\ t^{1/2}\ dt\right|^m d\lambda\right\}^{1/m}$$

$$\lesssim C \sup_{\ell} \sum_{k=0}^{1} \left\{\int_0^{\infty} \left|\Delta_t^{sk}(\ell)\overset{i}{\tilde{\phi}}(t,p)\right|^m t^{(1/2-\delta)m}(1+t)^{\delta m}\ dt\right\}^{\frac{1}{m}} |u|^{\sigma'},\ \delta > \frac{1}{m},$$

$$\sigma < \sigma' < s-\delta. \qquad \text{(II.42)}$$

Integrating this inequality to the m-th power with respect to $\overset{i}{p}$ and using the estimates (II.35) and

$$|u| < |h|\ (1 + |\lambda|)^{1/2} \qquad \text{(II.43)}$$

we obtain

$$\left[\int \left| \int_{-\infty}^{\infty} \frac{\Delta_t(-u)\ \tilde{\phi}^{i}(p,t)}{t + \tau^{ij}(p) - z}\ t_+^{1/2}\ dt \right|^m \ dp\right]^{\frac{1}{m}} \leq C(1 + |z|)^{\frac{\sigma'}{2}}\ |h|^{\sigma'} |v_\alpha|_{\tau+\sigma} |f|_{\alpha}{}_{H_\sigma} ,$$

where $\sigma' > \sigma$.

We now estimate the second integral on the right of (II.41). The following operators are familiar from the theory of spaces of differentiable functions:

$$(J^\gamma f)(\xi) = \int e^{i\xi\eta}\ \hat{f}(\eta)(1 + |\eta|^2)^{-\gamma/2} d\eta,\ \hat{f}(\eta) = \int e^{-i\eta\xi}\ f(\xi)\ d\xi. \tag{II.44}$$

Lemma II.3. Let $\Phi(k) = \phi(k,k)$, $\Phi'(k,k') \in S(R^6)$. Then

$$|\Phi|_{L_m(R^3)} < C|J_{k'}^{-\gamma}\ \Phi'|_{L_m(R^6)},\qquad \gamma > 3/m. \tag{II.45}$$

Proof. Let

$$G_\gamma(k) = \int e^{ikx}\ (1 + |x|^2)^{-\gamma/2}\ dx. \tag{II.46}$$

We have $\Phi' = J_{k'}^{\gamma}\ J_{k'}^{-\gamma}\Phi''$ and

$$\int |\Phi'(k,k)|^m\ dk = \int \left| \int G_\gamma(k - k')(J_{k'}^{-\gamma}\ \Phi')(k,k')\ dk' \right|^m dk$$

$$\leq \left\{\int |G_\gamma(k)|^{m'}\ dk\right\}^{m/m'} \int |(J_{k'}^{-\gamma}\Phi')(k,k')|^m\ dk\ dk',$$

where $\frac{1}{m} + \frac{1}{m'} = 1$. In view of the inequality $\int |G_\gamma(k)|^{m/(m-1)}\ dk < \infty$, $\gamma > 3/m$, this implies inequality (II.45), proving the lemma.

Using this lemma, the embedding theorem for Liouville and Sobolev spaces (see (S.2)) and inequality (II.40), we obtain an estimate for the second integral on the right of (II.41):

$$\int \left| J_{p_j}^{-\nu} \int_{-\infty}^{\infty} \frac{\tilde{\phi}^{i}(t-u,\ p)}{t+\tau^{ij}(p)-z} \ \frac{(t-u)_+^{1/2} - t_+^{1/2}}{|h|^{\sigma'}}\, dt \right|^m dp^i$$

$$\leq \int \left| \int_{-\infty}^{\infty} \frac{\left|J_{p_j}^{-\nu} J_{p_\ell}^{-\nu}\ \tilde{\phi}^{i\ell}(t-u,\ p_\ell',\ p)\right.}{t+\tau^{ij}(p)-z} \ \frac{(t-u)_+^{1/2} - t_+^{1/2}}{|h|^{\sigma'}}\, dt \right|^m dp^i\, dp_\ell'$$

$$\leq \sup_x \int \left| J_{p_j}^{-\nu} \quad J_{p_\ell}^{-\nu}\ \tilde{\phi}^{i}(x,p) \right|^m dp^i \ \sup_{p^i} \int_{-\infty}^{\infty} \left(\frac{\left|(t-u)_+^{1/2} - t_+^{1/2}\right|}{|h|^{\sigma'}}\right)^m dt$$

$$\leq |f|_{H_\sigma^\alpha}^m \ (1+|z|)^{\frac{\sigma''}{2}m} \ . \qquad \text{(II.47)}$$

Inequality (II.18) for $\alpha = \beta$ now follows from (II.37)--(II.41), (II.42), (II.47). This completes the proof of inequalities (II.17) and (II.18) for the case $\alpha = \beta$.

b) $\beta \neq \alpha$, $\alpha \cap \beta \neq \emptyset$. Let $\beta = (i,\ell)$. Set

$$\overset{i}{\phi}(q,p) = v_\alpha(p_j - q)\ (P_i f)(q, \overset{ij}{p}) \ . \tag{II.48}$$

We have

$$P_i g_\alpha(z,f) = \int \frac{\overset{i}{\phi}(q,p)\,dq}{(q + 1/2 \sum\limits_{s\neq i,j} p_s)^2 + \tau(\overset{ij}{p}) - z} \ . \tag{II.49}$$

By (II.49),

$$P_i g_\alpha(z,f) = \int_{-\infty}^{\infty} \frac{ds}{s + \tau(\overset{ij}{p}) - z} \int \overset{i}{\phi}(q,p)\,\delta\,[(q + 1/2 \sum_{k\neq i,j} p_k)^2 - s]\,dq \ . \tag{II.50}$$

The function (II.48) satisfies the estimate

$$\int | J_{p_\ell}^{-\nu} \ \phi(q) \ |^m_{\mathcal{H}_\alpha} \ dq < |v_\alpha|^m_\tau \, | \ f \ |^m_{\mathcal{H}_\alpha} \tag{II.51}$$

Consider the integral

$$\overset{i}{\phi}(p;\ s) = \int \overset{i}{\phi}(q,p)\,\delta\ (q + 1/2 \sum_{k\neq i,j} p_k)^2 - s \ \ dq. \tag{II.52}$$

Lemma II.4. Let $f(p) \in L_m(R^3)$. Then the function

$$(K_s f)(p;\ s) = \int f(q)\delta[(q-p)^2 - s]\, dq \tag{II.53}$$

satisfies the estimates

$$|\Delta_s^\theta(h)\Delta_s^\zeta(\ell)K_s f|_{L_m(R^3)} < Cs^\eta\,(1+s)^{\theta(1-\frac{2}{n})}\,(|f|_{L_m(R^3)} + \sup_u|\Delta^\gamma(h)f|_{L_m(R^3)}), \tag{II.54}$$

where $\theta + \zeta < \gamma + 2/m$, $\eta = \frac{1}{2}(1 - \frac{2}{m}) - \theta(1 - \frac{1}{m})$.

Proof. Consider the family of operators

$$T_z = J^{-z+\theta+\delta}\,\Delta_s^\theta(h)K_s \quad (J^\gamma = J_p^\gamma)\ , \tag{II.55}$$

where $\delta > 0$, $1 \geq \theta \geq 0$ are arbitrary but fixed numbers. Note that K_s is a convolution-type operator admitting the representation

$$(K_s f)(p;s) = \int e^{ipx}\,\hat{f}(x)\,\frac{\sin\sqrt{s}|x|}{|x|}\,dx,\quad \hat{f}(x) = \int e^{-ixp} f(p)dp. \tag{II.56}$$

It follows from (II.56) and the Plancherel theorem that the family of operators T_z is weakly analytic with respect to z, $0 \leq \operatorname{Re} z \leq 1$, in $L_2(R^3)$ and hence also in the class of simple functions on R^3 (i.e., functions with compact support expressible as finite linear combinations of characteristic functions of squared beams in R^3).

To derive estimates for the operator T_z, we proceed as follows:

$$\begin{aligned}|T_{iy}f|_{L_\infty(R^3)} &\leq |J^{-iy+\delta}\,\Delta_s^\theta(h)\,K_s J^\theta f|_{L_\infty(R^3)}\\ &\leq C\sup_p\left|\int \Delta_s^\theta(h)\,[\sqrt{s}(J^\theta f)(\sqrt{s}\omega + p)]d\omega\right|\\ &\leq C|f|_{L_\infty(R^3)}(s^{1/2-\theta} + s^{1/2})\ .\end{aligned} \tag{II.57}$$

Using (II.56), we obtain

$$\|T_{1+iy}\|^2_{L_2(\mathbb{R}^3)} = C \int |\hat{f}(x)|^2 \frac{|\sin\sqrt{s+h}|x| - \sin\sqrt{s}|x||^2}{|h|^{2\theta}|x|^2} (1+|x|^2)^{1-\theta-\delta} dx$$

$$\leqslant C\, s^{-\theta} \|f\|^2_{L_2(\mathbb{R}^3)} . \tag{II.58}$$

The estimates (II.57), (II.58) and the interpolation theorem for families of operators ([18], Theorem 1.11) imply the estimate

$$\|T_{2/m} f\|_{L_m(\mathbb{R}^3)} \leqslant C\, s^{\eta} (1+s)^{\theta(1-2/m)} \|f\|_{L_m(\mathbb{R}^3)} . \tag{II.59}$$

Using the estimate (II.59), the definition (II.55) of T_z, embedding theorem for Liouville classes of functions (see (S.2)) and the relation

$$[K_s, \Delta(h)] = 0,$$

we obtain (II.54), proving the lemma.

It follows from Lemma II.4 that the function

$$\overset{i}{\Phi}{}'(p,p';\, s) = \int \overset{i}{\phi}(q,p)\delta[(q + 1/2 \sum_{k\neq i,j,\ell} p_k + 1/2p')^2 - s]\, dq$$

satisfies the estimate

$$\int |\, J^{-\nu}_{p_\ell} J^{-\nu}_{p_j} \Delta^{\theta}_s(v) \overset{i}{\Phi}{}'(p,p';\, s)|^m\, dp'\, d\overset{i}{p}$$

$$\leqslant C\, s^{\eta_1 m} (1+s)^{\theta(1-\frac{2}{m})m} \int\int |\, J^{-\nu}_{p_\ell} J^{-\nu}_{p_j}\, \overset{i}{\phi}(q,p)|^m\, dq\, d\overset{i}{p}$$

$$\leq C\, s^{\eta_1 m}\, (1+s)^{\theta(1-\frac{2}{m})m}\, |v_\alpha|_\nu^m\, |f|_{\overset{\alpha}{H}}^m\,, \qquad \theta < \frac{2}{m}\,. \tag{II.61}$$

By the definition of the function $\overset{i}{\Phi}(p;s)$,

$$\overset{i}{\Phi}(p;\ s) = \overset{i}{\Phi}{}'(p,p_\ell;\ s)\,. \tag{II.62}$$

The estimate (II.61), Eq. (II.62) and Lemma II.3 imply an estimate for $\overset{i}{\Phi}(p;\ s)$:

$$|\Delta_s^\theta(h)\Phi(s)|_{\overset{\alpha}{H}} \leq C\, s^{\eta_1}\, (1+s)^{\theta(1-\frac{2}{m})}\, |v_\alpha|\, |f|_{\overset{\beta}{H}}\,, \tag{II.63}$$

where $0 \leq \theta < \frac{2}{m}$. Hence the function

$$P_i g_\alpha(z;f) = \int_{-\infty}^{\infty} \frac{\overset{i}{\Phi}(p;\ s)\, ds}{s + \overset{ij}{\tau}(p) - z} \tag{II.64}$$

satisfies the inequality

$$|\Delta_\lambda^\theta(h) g_\alpha(z,f)|_{\overset{\alpha}{H}} \leq C|v_\alpha|_\tau\, |f|_{\overset{\alpha}{H}}\,, \quad 0 \leq \theta < \frac{2}{m} \tag{II.65}$$

We can now prove estimate (II.18) for this case. Let $f \in \overset{\beta}{H}_\sigma$. Estimating the functions $\Delta_{p_k}^\sigma(h)\overset{i}{\Phi}(p;\ s)$ and $\Delta^\sigma{}_s(h)\overset{i}{\Phi}(p;\ s)$ in the same way as $\overset{i}{\Phi}(p;\ s)$, we obtain the inequalities

$$|\Delta_s^\theta(h)\Phi(s)|_{H_\rho^\alpha} < C\, s^{\eta_1}(1+s)^{\theta(1-\frac{2}{m})} |v_\alpha|_{\tau+\sigma}|f|_{H_\sigma^\beta} ,$$

where $\theta + \rho < \sigma + \frac{2}{m}$ $\rho \leqslant \sigma$. The integral (II.64) is then estimated in the same way as (II.23) (see above, beginning from Eq. (II.35)). This completes the proof of estimate (II.18) for $\alpha \neq \beta$, $\alpha \cap \beta \neq \emptyset$.

c) $\alpha \cap \beta = \emptyset$. Let $\beta = (t,\ell)$. We have

$$p_i g_\alpha(z,f) = \int \frac{v_\alpha(p_j + 1/2 \sum_{s\neq i,j} p_s - k)}{k^2 + \tau_1^{ij}(p) - z} (P_t f)(k - 1/2 \sum_{s\neq i,j} p_s ,$$

$$-k - 1/2 \sum_{s\neq i,j} p_s, \overset{ijt}{p})dk . \tag{II.66}$$

Our estimates of this function will be derived from estimates for the function obtained from it by the substitution

$$p_t \to 2p_t - \sum_{s\neq i,j,t} p_s . \tag{II.67}$$

Introduce the notation

$$\overset{\alpha}{\tau}'(p) = \overset{\alpha}{\tau}(p)\Big|_{p_t \to 2p_t - \sum_{k\neq i,j,t} p_k} . \tag{II.68}$$

We now proceed to estimate the integral (II.66). Set

$$\overset{i}{\phi}(p,k) = v_\alpha(p_j + p_t - k)(p_t f)(k - p_t, -k-p_t, \overset{ijt}{p}) . \tag{II.69}$$

Let $f \in \overset{\beta}{H}$. We have the following estimate for the function (II.69):

$$\int |J^{-\nu}_{p_\ell}\, \phi(k)|^{m}_{\underset{H}{\alpha}}\, dk \leqslant C\, |v_\alpha|^{m}_{\tau}\, |f|^{m}_{\underset{H}{\beta}} \tag{II.70}$$

Consider the function

$$\int \overset{i}{\phi}(p,\ \sqrt{t}\ \omega)\, d\omega, \quad t \geqslant 0. \tag{II.71}$$

Continue this function with respect to t to the interval $(-\infty, 0)$ as a smooth function decreasing rapidly to $-\infty$. Denote the extended function by $\overset{i}{\tilde{\phi}}(p,t)$. In terms of this function, the integral (II.66) is written

$$(P_i\, g_\alpha(z,f))\overset{i}{(p;\ z)} = \overset{i}{g}'(p;\ z)\Big|_{p_t \to 1/2 \sum_{s \neq i,j} p_s}\ , \tag{II.72}$$

where

$$\overset{i}{g}'(p(p;\ z) = \int \frac{\overset{i}{\tilde{\phi}}(p,s)\, s_+^{1/2}}{s + \overset{ij}{\tau}'(p) - z}\, ds\ . \tag{II.73}$$

We obtain the following estimates from (II.70), (II.71):

$$\int_{-\infty}^{\infty} |J^{-\nu}_{p_\ell}\, \tilde{\phi}(s)|^{m}_{\underset{H}{\alpha}}\, |s|^{1/2}(1 + |s|)^{\frac{m}{2} - 1}\, ds \leqslant C\, |v_\alpha|^{m}_{\tau}\, |f|^{m}_{\underset{H}{\beta}} \tag{II.74}$$

Using Lemma II.3 and estimate (II.74) for the integral (II.73), we obtain the inequalities

$$\int \Big|J^{-\nu}_{p_j} \int_{-\infty}^{\infty} \frac{\overset{i}{\tilde{\phi}}(p,s)}{s + \overset{ij}{\tau}'(p) - z}\, s_+^{1/2}\, ds\Big|^{m}\, \overset{i}{dp}$$

$$\leqslant \int_{\infty}^{\infty} \Big|J^{-\nu}_{p_j}\, J^{-\nu}_{p'} \int_{-\infty}^{\infty} \frac{\overset{i\ell}{\tilde{\phi}}(p,p',s)}{s + \overset{ij}{\tau}'(p) - z}\, s_+^{1/2}\, ds\Big|^{m}\, \overset{i}{dp}\, dp'$$

$$\leq \int \int_{-\infty}^{\infty} |J_{p_j}^{-\nu} J_{p_\ell}^{-\nu} \tilde{\phi}(\overset{i}{p}, s)|^m s_+^{m/2} \, ds \, d\overset{i}{p}$$

$$\leq C|v_\alpha|_\tau^m |f|_{\substack{\beta \\ H}}^m \tag{II.75}$$

Thus

$$|g'(z)|_{\substack{\alpha \\ H}} \leq C|v_\alpha|_\tau |f|_{\substack{\beta \\ H}} \tag{II.76}$$

Eqs. (II.72) and (II.76) now imply the following estimate for (II.66):

$$|\overset{i}{g}(z; f)|_{\substack{\alpha \\ H}} \leq C|v_\alpha|_\tau |f|_{\substack{\beta \\ H_\lambda}} \tag{II.77}$$

We can now derive the estimate (II.18) for this case $(\alpha\cap\beta = \emptyset)$. Let $f \in H^{\beta}_{\sigma,\lambda}$. Lemma II.2 implies the following estimate for $\tilde{\phi}^{i}(p,s)$:

$$\int\int_{-\infty}^{\infty} |\Delta_s^{\sigma}(h) J_{p_\ell}^{-\nu} \tilde{\phi}(s)|_{H^{m}_{\alpha}} |s|^{1/2}(1+|s|)^{\frac{m}{2}-1} ds \frac{dh}{|h|}$$

$$\leq C |v_\alpha|_{\tau+\sigma}^{m} |f|_{H^{\beta}_{\sigma}} \tag{II.78}$$

To estimate the integral (II.75) in this case, we use in the main the same techniques as before. We shall therefore cite only the main inequalities, sketching the rest of the proof. Let $k \neq i,j$ and

$$v = \tau'^{ij}(p + he_k) = \tau'^{ij}(p) = a(p)h + bh^2, \quad b > 0. \tag{II.79}$$

By (II.73)

$$\Delta^{\sigma}_{p_k}(h) g'^{i}(p;z) = \int_{-\infty}^{\infty} \frac{\Delta^{\sigma}_{p_k}(h)\tilde{\phi}^{i}(p,s)}{s + \tau'^{ij}(p) - z} s_+^{1/2} ds +$$

$$+ |h|^{-\sigma} \{ \int_{-\infty}^{\infty} \frac{\tilde{\phi}^{i}(p,s) s_+^{1/2} ds}{s + \tau'^{ij}(p + he_k) - z} - \int_{-\infty}^{\infty} \frac{\tilde{\phi}^{i}(p,s) s_+^{1/2} ds}{s + \tau'^{ij}(p) - z} \tag{II.80}$$

An estimate for the first term on the right of this equality follows at once from inequality (II.76). To estimate the second integral on the right of (II.80) we proceed as in the previous derivations of inequality (II.18), distinguishing between two cases: $\tau'^{ij}(p) > 10(1 + |\lambda|)$ and $\tau'^{ij}(p) < 10(1 + |\lambda|)$. In the first case we use the fact that for $|h| < 1$ the integrals in braces on the right of (II.80) are nonsingular. In the second case we apply a suitable substitution in the first integral in braces, to reduce the second term on the right of (II.80) to the form

$$\int_{-\infty}^{\infty} \frac{\Delta_s(-v)\tilde{\phi}^{i}(p,s)}{s + \tau'_{ij}(p) - s} s_+^{1/2}\, ds + \int_{-\infty}^{\infty} \frac{\phi^{i}(p,s - v)}{s + \tau'_{ij}(p) - z} [(s - v)_+^{1/2} - s_+^{1/2}]\, ds. \tag{II.81}$$

The first integral in (II.81) is estimated in the same way as (II.73), except that one also needs inequalities (II.43), (II.78) and the inequality

$$\int_0^\infty\int_0^\infty |\Delta_s(-v) J_{p_j}^{-\nu} J_{p_\ell}^{-\nu} \tilde{\phi}^{i}(p,s)|^m s_+^{m/2}\, ds \frac{dh}{|h|^{1+\sigma m}}$$

$$\lesssim \int_0^\infty J_{p_j}^{-\nu} J_{p_\ell}^{-\nu} \tilde{\phi}^{i}(p,s)|^m s^{1/2}\, ds +$$

$$+ (1 + |\lambda|)^{\sigma m} \int\int_0^\infty |\Delta_s^\sigma(h) J_{p_j}^{-\nu} J_{p_\ell}^{-\nu} \tilde{\phi}^{i}(p,s)|^m s^{m/2}\, ds \frac{dh}{|h|} . \tag{II.82}$$

This inequality will be proved below.

Finally, the second integral in (II.81) is estimated in the same way as the last term in (II.41).

It remains to prove inequality (II.82). Let $f \in S(R^1)$ and $|a| > 0$. We shall find estimates for certain integrals involving $f(x)$. We have

$$\int_{-\infty}^\infty\int_0^\infty |\Delta(ah - h^2) f(x)|^m x^{m/2}\, dx \frac{dh}{|h|^{1+\sigma m}}$$

$$\lesssim C \int_{-\infty}^\infty\int_{ah}^\infty |\Delta(-h^2) f(y)|^m (y - ah)^{\frac{m}{2}}\, dy \frac{dh}{|h|^{\sigma m+1}} +$$

$$+ \int_{-\infty}^{\infty}\int_{0}^{\infty} |\Delta(ah)f(x)|^m x^{\frac{m}{2}} dx \frac{dh}{|h|^{\sigma m+1}} \leq$$

$$\leq C|a|^{\frac{m}{2}} \int_{-\infty}^{\infty}\int_{ah}^{\infty} |\Delta(-h^2)f(y)|^m dy \frac{dh}{|h|^{1+(\sigma-1/2)m}} +$$

$$+ \int_{-\infty}^{\infty}\int_{0}^{\infty} |\Delta(-h^2)f(y)|^m y^{\frac{m}{2}} dy \frac{dh}{|h|^{1+\sigma m}} +$$

$$+ C|a|^{\sigma m} \int_{-\infty}^{\infty}\int_{0}^{\infty} |\Delta^{\sigma}(h)f(y)|^m y^{m/2} dy \frac{dh}{|h|} \leq$$

$$\leq C|a|^{\frac{m}{2}} \int_{-\infty}^{\infty}\int_{-\infty}^{\infty} |\Delta^{1/2(\sigma-1/2)}(\ell)f(x)|^m dx \frac{d\ell}{|\ell|} +$$

$$+ C|a|^{\sigma m} \int_{-\infty}^{\infty}\int_{0}^{\infty} |\Delta^{\sigma}(\ell)f(x)|^m x^{m/2} dx \frac{d\ell}{|\ell|} .$$

Using this inequality, Eqs. (II.15), (II.79) and the condition $\tau'^{ij}(p) < 10(1 + |\lambda|)$, we obtain the estimate (II.82).

This completes the proof of inequalities (II.17) and (II.18).

APPENDIX III

In this appendix we study singular integrals of type (6.50), subject to certain conditions to be stated below.

Since $R^i \geq 0$, $\sum R^i > 0$, there exists $M(z)$ such that for $|x_j| > M(z)$ the denominators involving x_j are nonsingular. It follows that for sufficiently large values of any specific selection of variables from (p,q) the function $Q(p,q;z)$ is differentiable arbitrarily many times with respect to them. If we include the denominators containing these variables in the function Φ, our problem reduces to the investigation of integrals of type (6.50) in which some of the external variables do not enter into the denominator of the integrand and those that do vary in a bounded region whose size depends on $|z|$.

Let $\mathfrak{P}_N$ denote the set of N-dimensional matrices R satisfying the conditions:

1) $R = \overline{R}$, $R = R^T$.

2) The matrix remaining after zero rows and columns are deleted from R is positive defined.

The set $\mathfrak{P}_N$ is a linear space over the positive numbers. Matrices of $\mathfrak{P}_N$ may be regarded as operators in the space R^{3N}, whose vectors are written $p = (p_1,\ldots,p_N)$, $p_i \in R^3$, $i = 1,\ldots,N$, according to the formúla

$$(Rp)_i = \sum_{j=1}^{N} (R)_{ij}\, p_j, \quad i = 1,\ldots,N.$$

Lemma III.1. Let $\Phi(k,v)$, $k \in R^{3m}$, $v \in R^{3\ell}$, be a function such that for all α

$$|(D^{\alpha}\Phi)(k,v)| < C\, N(v)\, N'(k,v),$$

where $N(v)$, $N'(k,v)$ are certain estimating functions, $\int N'(k,v)dk = C < \infty$; let R^i, $i = 1,\dots,s$, be matrices in $\mathfrak{P}_N$ such that

$$\sum_{i=1}^{s} R^i > 0;$$

for all $j_1,\dots,j_t$, $t \leq s$, there exist $i_1,\dots,i_k$, $i_\ell \leq m$, $3k/2 > t$, such that

$$(R^{j_{q(p)}})_{i_p i_p} > 0, \quad p = 1,\dots,k. \tag{III.2}$$

Then the integral

$$J(u,v;\, z_1,\dots,z_s) = \int \frac{\Phi(k,v)\, d^m k}{\prod_1^s [(x,R^i x) - z_i]}, \quad x = (k,u),\ u \in R^{3n}, \tag{III.3}$$

satisfies the estimates

$$|\Delta_u^t(h)\, D_v^k\, J(u,v;\, z_1,\dots,z_s)$$

$$< CN(v) \sum_{i=1}^{s} |\Delta_i(u) - \operatorname{Re} z_i|^{-\varepsilon} \ln(1 + \sum_{i=1}^{s} |\operatorname{Re} z_i|)$$

for all k, $0 \leq t < 1$, $\varepsilon = \theta(t - 1/2)(t - 1/2 + \delta)$, $\delta > 0$, where

$$\Delta_i(u) = (u,\, T^i u),\ T^i = C^i - \lim_{\varepsilon\to 0}(B^i)^T(A^i + \varepsilon E)^{-1}B^i,$$

$$A^i = |R^i_{k,j},\ k = 1,\dots,m;\ \ j = 1,\dots,m|\ ,$$

$$B^i = |R^i_{k,j},\ k = m+1,\dots,m+n;\ \ j = 1,\dots,m|\ ,$$

$$C^i = |R^i_{k,j},\ k = m+1,\dots,m+n;\ \ j = m+1,\dots,m+n|\ .$$

Proof. To avoid encumbering the formulas, we set $z_1 = \dots = z_s = z$, $J(u,v;\, z) = J(u,v;\, z,\dots,z)$. Using the Feynman identity

$$\prod_1^s A_i^{-1} = \int_o^1 \frac{\delta(1 - \sum_1^s \alpha_i)}{(\sum_1^s \alpha_i A_i)^s} d^s\alpha ,$$

we transform the integral (III.3) to the form

$$J(u,v;\ z) = \int_o^1 \int \frac{\Phi(k,v)\delta(1 - \sum_i^s \alpha_i)}{[\omega(k,u,\alpha) - z]^s} d^m k\, d^s\alpha , \tag{III.4}$$

where

$$\omega(k,u,\alpha) = (k,A(\alpha)k) + 2(k,B(\alpha)u) + (u,c(\alpha)u) ,$$

$$A(\alpha) = \sum_i \alpha_i A^i,\ B(\alpha) = \sum_i \alpha_i B^i,\ C(\alpha) = \sum_i \alpha_i C^i .$$

Let $V(\alpha)$ be a unitary matrix diagonalizing $A(\alpha)$. The substitution

$$k = v(\alpha)h - A^{-1}(\alpha)B(\alpha)u$$

brings the integral (III.4) to the form

$$J(u,v;\ z) = \int \frac{\psi(h,u,v,\alpha)\delta(1 - \sum_1^s \alpha_i)}{[\sum_1^n \lambda_i(\alpha)h_i^2 + \Delta(u,\alpha) - z]^s} d^m h\, d^s\alpha , \tag{III.5}$$

where $\lambda_i(\alpha)$ are the eigenvalues of $A(\alpha)$,

$$\psi(h,u,v,\alpha) = \Phi(V(\alpha)h - A^{-1}(\alpha)B(\alpha)u,v),$$

$$\Delta(u,\alpha) = (u,C(\alpha)u) = (u,B^T(\alpha)A^{-1}(\alpha)B(\alpha\ )u). \tag{III.6}$$

It follows from the properties of the matrices R^i that $\operatorname{Ker} A(\alpha) \subset R^m$ $\operatorname{Im} B(\alpha)$ for $\alpha \in \{\alpha = (\alpha_1,\ldots,\alpha_s),\ 0 \leq \alpha_i \leq 1,\ \sum_1^s \alpha_i = 1.$ Hence the inequality

$$|A^{-1}(\alpha)B(\alpha)| < C. \tag{III.7}$$

Formulas (III.6), (III.7) now imply estimates for the function $\psi(h,u,v,\alpha)$:

$$|(D^\beta\psi)(h,u,v,\alpha)| < CN(v)N'(V(\alpha)h - A^{-1}(\alpha)B(\alpha)u,v). \tag{III.8}$$

Integrating in (III.5) with respect to the angles $\omega(h_1),\ldots,\omega(h_m)$ and then substituting $h_i^2 = t_i$, we obtain

$$J(u,v;\ z) = \int_0^1\int_0^\infty \frac{\tilde{\psi}(t,u,v,\alpha)\prod_1^m t_i^{1/2}\, d^m t}{[\lambda(\alpha)t + \Delta(u,\alpha) - z]^s}\,\delta(1 - \sum_1^s \alpha_i)\, d^s\alpha\ , \qquad \text{(III.9)}$$

where $\lambda(\alpha)t = \sum_1^m \lambda_i(\alpha)t_i$,

$$\tilde{\psi}(h_1^2\ldots h_m^2,u,v,\alpha) = \int \psi(h_1\ldots h_m,u,v,\alpha)\prod_1^m d\omega(h_i).$$

The function $\tilde{\psi}(t,u,v,\alpha)$ satisfies the estimates

$$|(D_t^\beta D_u^\gamma D_v^\delta\tilde{\psi})(t,u,v,\alpha)| < C\ N(v)\ N_1(t,u,v,\alpha) \qquad \text{(III.10)}$$

for all β,γ,δ, where

$$N_1(h_1^2\ldots h_m^2,u,v,\alpha) = \int N'(V(\alpha)h - A^{-1}(\alpha)B(\alpha)u,v)\prod_1^m d\omega(h_1).$$

Next, we have

$$J(u,v;\ z) = \int_0^1\int_0^\infty \frac{d\rho}{(\rho + \Delta(u,\alpha)-z)^s}\int_0^\infty \tilde{\psi}(t,u,v,\alpha)\,\delta(\lambda t-\rho)\ \times$$

$$\times \prod_1^m t_i^{1/2}\ d^m t\ \delta(1 - \sum_1^s \alpha_i)\ d^s\alpha\ . \qquad \text{(III.11)}$$

Consider the function

$$F(\rho,u,v,\alpha) = \int_0^\infty \tilde{\psi}(t,u,v,\alpha)\,\delta(\lambda t-\rho)\prod_1^m t_i^{1/2}\ d^m t\ .$$

Let us estimate this function. By the assumptions of the lemma, $A(\alpha) > 0$ for $\alpha_j > 0$, $j = 1,\ldots,s$, and so also $\lambda_i(\alpha) > 0$ $(\alpha_j > 0,\ j = 1,\ldots,s)$,

$i = 1,\dots,m$. For $k_i = k_i(\alpha) < 3/2$, $i = 1,\dots,m, \sum k_i = k$, we have

$$(D_\rho^k F)(\rho,u,v,\alpha) = \int_0^\infty \tilde{\psi}(t,u,v,\alpha)\delta^{(k)}(\lambda(\alpha)t - \rho) \prod_1^m t_i^{1/2} d^m t$$

$$= \prod_{i=1}^m \lambda_i^{-k_i}(\alpha) \int_0^\infty \tilde{\psi}(t,u,v,\alpha)(\prod_1^m D_{t_i}^{k_i} \delta(\lambda(\alpha)t-\rho))\prod_1^m t_i^{1/2} d^m t$$

$$= \prod_{i=1}^m \lambda_i^{-k_i}(\alpha) \int_0^\infty \delta(\lambda(\alpha)t-\rho) \prod_1^m \frac{t_{i-}^{-1-k_i}}{\Gamma(-k_i)} * (\psi(t,u,v,\alpha) \prod_1^m t_i^{1/2})\, d^m t\ , \qquad \text{(III.12)}$$

where $f(t)*g(t)$ denotes the convolution of functions (usua-ly generalized functions). (For the definition of the generalized function $t_-^{-\lambda}$ (see [38]).

We now divide the domain of variation of α into subregions

$$Q_1 = \{\alpha = (\alpha_1,\dots,\alpha_s) \in [0,1]^s,\ \alpha_1 > 1/2\, s^{-1}\}\ ,$$

$$Q_i = \{\alpha = (\alpha_1,\dots,\alpha_s) \in [0,1]^s \setminus \bigcup_1^{i=1} Q_j,\ \alpha_i > 1/2\, s^{-1}\},\ i = 2,\dots,s,$$

$$Q_o = [0,1]^s \setminus \bigcup_1^s Q_i = \{\alpha = (\alpha_1,\dots,\alpha_s) \in [0,1]^s,\ \alpha_j \leqslant 1/2\, s^{-1},\ j = 1,\dots,s\}.$$

We shall estimate the function (III.12) separately in each region Q_i, taking a special set $\{k_j\}$ for each Q_i. Since $A^i \neq 0$, $A^i \geqslant 0$, $i = 1,\dots,s$, it follows that for any i there exists n_i such that $D_{\alpha_i} \lambda_{n_i}(\alpha) > 0$. Let $\alpha \in Q_i$. Then $\alpha_i > 1/2\, s^{-1}$, whence it follows that

$$\lambda_{n_i}(\alpha) > C\, s^{-1},\quad C > 0.$$

Now, successively using the inequality

$$|t_-^{-1-k} * (\psi(t)t^{1/2})| < M(t)t^{1/2-k} \sup_x N^{-1}(x)\,|\psi(x)| + t^{1/2}|t_-^{-1-k}*\psi(t)|$$

$$< (M(t) + N(t))(t^{1/2-k} + t^{1/2}) \sup_x N^{-1}(x)(|\psi(x)| + |x_-^{-1-k}*\psi(x)|),$$

where

$$M(t) = \int_0^\infty \frac{N(tx)\,dx}{x^{1/2}(x-1)^k} \quad (\int M(t)t^\alpha\,dt = C\int N(t)t^\alpha\,dt,\ \alpha \geq 0),$$

integrating in (III.12) with respect to t_{n_i}, we obtain estimates for $D_\rho^k F$:

$$|(D_\rho^k D_v^p F)(\rho,u,v,\alpha)| < \rho^{-\varepsilon} \prod_1^m \lambda_j^{-k_j}(\alpha)\ N(v)\ N_2(\rho,u,v,\alpha)\ , \tag{III.13}$$

where $\sum_1^m k_i = k$, $k_j < 3/2$, $\rho^\varepsilon D_\rho^k D_v^p F|_{\rho=0} = 0$, $\varepsilon = \max(0, k_{n_i} - 1/2)$,

$|p| \geq 0$, $\alpha \in Q_i$ $(i = 1,\ldots,s)$, and $N_2(\rho,u,v,\alpha)$ is some estimating function such that

$$\int N_2(t,u,v,\alpha)d\rho < C\int N'(k,v)\ dk < C. \tag{III.14}$$

Integrating by parts in (III.11), we obtain

$$J(u,v,z) = \int_0^1 \frac{D_\rho^{s-1}F(\rho,u,v,\alpha)}{\rho + \Delta(u,\alpha) - z}\, d\rho\,\delta(1 - \sum_1^s \alpha_i)\ d^s\alpha\ . \tag{III.15}$$

Dividing the integration domain with respect to α into the subregions Q_i, $i = 1,\ldots,s$, and noting that the integral over Q_o vanishes because of the factor $\delta(1 - \sum_1^s \alpha_i)$ in the integrand, we deduce the following estimates from (III.14) $(\sum_{j\neq n_i} k_j = k - k_{n_i} = s - 1 + \mu+\nu - (k_{n_i} - 1/2) - 1/2 > s - 1 + \mu+\nu - 1/2 - \varepsilon)$:

$$|\Delta_u^\mu(h)\Delta_z^\nu(\ell)\ D_v^p\ J(u,v;\ z)|$$

$$\leq C \sum_{i=1}^s \int_{Q_i} |\Delta(u,\alpha) - \operatorname{Re} z|^{-\varepsilon} \prod_{j\neq n_i} \lambda_j^{-k_j}(\alpha)\delta(1 - \sum_1^s \alpha_j)\ d^s\alpha\cdot N(v)\ , \tag{III.16}$$

where $\varepsilon \geq 0, \forall |p| \geq 0$. Since $\lambda_i(0) = 0, \lambda_i(\alpha) > 0$ $(\alpha > 0)$ and $\Delta(u,\alpha)$ is a polynomial in α, we have

$$\int \prod_1^m \lambda_i^{-k_i}(\alpha) |\Delta(u,\alpha)-\lambda|^{-\varepsilon} \delta(1 - \sum_1^s \alpha_i) \, d^s\alpha$$

$$\leq |\Delta(u,\alpha^i)-\lambda|^{-\varepsilon} \int \prod_1^m \lambda_j^{-k_j}(\alpha) \delta(1 - \sum \alpha_i) \, d^s\alpha , \quad (\alpha^i)_k = \delta_{k_i} \tag{III.17}$$

We must show that the integrals on the right of inequality (III.17) are finite; in other words, we consider integrals of the type

$$\int_{Q_i} \prod_{j \neq n_i} \lambda_j^{-k_j}(\alpha) \delta(1 - \sum_1^s \alpha_i) \, d^s\alpha , \tag{III.18}$$

where $\sum_{j \neq n_i} k_j = s - 1 - \varepsilon_1, \quad \varepsilon_1 > 0.$

Let $\lambda_i(A)$, $i = 1,\ldots,m$, denote the eigenvalues of a matrix $A \in \mathfrak{P}_m$. indexed in order of their appearance in the diagonalization of A.

Lemma III.2. Let $A(\alpha) = \sum_1^s \alpha_i A^i$, $A^i \in \mathfrak{P}_m$, $0 \leq \alpha_i \leq 1$, $i = 1,\ldots,s$, $\sum \alpha_i = 1$. There exists a number C such that

$$\lambda_i(A(\alpha)) > C[A(\alpha)]_{ii} .$$

Proof of Lemma III.2. It will suffice to show that for all $i_1,\ldots,i_k$, $k \leq s - 1$, $i = 1,\ldots,s$,

$$\lim_{\substack{\alpha_{i_1} \to 0 \\ \cdots \\ \alpha_k \to 0}} \frac{\lambda_i(A + \sum \alpha_j B_j)}{(A + \sum \alpha_j B_j)_{ii}} > C, \quad C > 0. \tag{III.20}$$

for this, in turn, we need only show that for $A, B_j \in \mathfrak{P}_m$, $j = 1,\ldots,k$,

$$\lim_{\substack{\alpha_1\to o\\ \cdots\\ \alpha_k\to o}} \frac{\lambda_i(A + \sum \alpha_j B_j)}{(A + \sum \alpha_j B_j))_{ii}} > C, \quad C > 0 . \tag{III.20}$$

If $(A)_{ii} > 0$, the proof of (III.20) is elementary:

$$\lim_{\substack{\alpha_1\to o\\ \cdots\\ \alpha_k\to o}} \frac{\lambda_i(A + \sum \alpha_j B_j)}{(A + \sum \alpha_j B_j)_{ii}} = \frac{\lambda_i(A)}{(A)_{ii}} > 0 .$$

Now let $(A)_{ii} = 0$. Let $\phi_i(\alpha)$ be an eigenvector of the matrix $A + \sum \alpha_j B_j$ belonging to the eigenvalue $\lambda_i(A + \sum \alpha_j B_j)$, so chosen that $\phi_i(0) = (\delta_{i,k}, k = 1,\ldots,m)$. For sufficiently small α,

$$\begin{aligned}
\lambda_i(A + \sum \alpha_j B_j) &= (\phi_i(\alpha), (A + \sum \alpha_j B_j)\phi_i(\alpha)) \\
&= (\phi_i(\alpha), A\phi_i(\alpha)) + \sum \alpha_j(\phi_i(\alpha), B_j\phi_i(\alpha)) \\
&> C \sum \alpha_j(\phi_i(0), B_j\phi_i(0)) \\
&= C \sum \alpha_j(B_j)_{ii} \\
&= C(A + \sum \alpha_j B_j)_{ii} .
\end{aligned} \tag{III.21}$$

The relation (III.20) follows from (III.21), proving Lemma III.2.

Lemma III.2 implies an estimate for the integrals (III.18):

$$\int_{Q_i} \prod_{j\neq n_i} \lambda_j^{-k_j}(\alpha)\delta(1 - \sum_1^s \alpha_k)d^s\alpha \leq \int_{Q_i} \prod_{j\neq n_i} [A(\alpha)]_{jj}^{-k_j} \delta(1 - \sum_1^s \alpha_k)d^s\alpha . \tag{III.22}$$

The factors $[A(\alpha)]_{jj}$ in the integrand on the right may be estimated with the help of the inequality

$$(\sum_k \alpha_{i_k})^{-1} \leqslant C \prod_k \alpha_{i_k}^{-r_k} , \quad \sum_k r_{i_k} = 1.$$

It follows from conditions (III.1), (III.2) that we can choose k_j, $j \neq n_i$, $k_j < 3/2$, $\sum_{j \neq n_i} k_j < s - 1$, in such a way that the integral on the right of (III.22) is finite. This completes the proof of Lemma III.1.

APPENDIX IV

Let $R^n = R^{n_1} \oplus R^{n_2}$, $n_1 + n_2 = n$, and J_1^δ be a J^δ-operator defined with respect to the variables in R^{n_1}. Let, furthermore, Ω be the sphere $|p| = 1$ in R^n and a point $p \in R^n$ with spherical coordinates $s = |p|$ and $\omega = |p|^{-1} p \in \Omega$ be written as $p = s\omega$. A function $f \in L_m(R^n)$ can be viewed as a vector function $r \to f(\sqrt{r}\cdot)$ from R^+ into $L_m(\Omega)$ with the corresponding embedding denoted by $\Pi(r)$:

$$(\Pi(r)f)(\omega) = f(\sqrt{r}\omega) \ .$$

We have

<u>Lemma IV.1.</u> The family $\Pi(r)$, $r > 0$, is bounded, uniformly in every compact subset of R^+, from $J_1^\nu L_m(R^n) (\equiv L_m^\nu(R^{n_1}) \otimes L_m(R^{n_2}))$ into $L_m(\Omega)$, $\nu > \frac{n_1}{m}$. Moreover, its δ-derivative, $\Delta_r^\delta(h)\Pi(r)$, $0 \leqslant \delta \leqslant 1$, is bounded from $J^\delta J_1^\nu L_m(R^n)$ into $L_m(\Omega)$:

$$|\Delta_r^\delta(h)\Pi(r)f|_{L_m(\Omega)} \leqslant Cr^{-\frac{1}{2m}(n_2+\delta m)} |f|_{J^\delta J_1^\nu L_m(R^n)} \tag{IV.1}$$

<u>Remark.</u> The statement, similar to the first part of this lemma, is contained in the paper cited in the footnote to p. 54 (Proposition (2.2)).

<u>Proof.</u> The boundedness of $\Pi(r)$ follows easily:

$$|\Pi(r)f|^m_{L_m(\Omega)} = r^{-\frac{n}{2}+1} \int |f(p)|^m \delta(p^2-r)dp =$$

$$= r^{-\frac{n}{2}+1} \int |f(p_1,p_2)|^m \ \delta(p_1^2+p_2^2-r)dp_1dp_2 =$$

$$= r^{-\frac{n}{2}+1} \int |f(\sqrt{r-p_2^2}\ \omega_1,p_2)|^m \ (r-p_2^2)_+^{\frac{n_1}{2}-1} \ d\omega_1 \ dp_2 \tag{IV.2}$$

Taking into account the following representation of $f \in J_1^{\nu} L_m(R^n)$ as

$$f(p) = (J_1^{\nu} J_1^{-\nu} f)(p) = \int G_{\nu}(p_1 - q_1)(J_1^{-\nu} f)(q_1, p_2) dq_1$$

and the property

$$|G_{\nu}|_{m'} < \infty \quad \text{for} \quad \nu > \frac{n_1}{m}, \quad m^{-1} + m'^{-1} = 1,$$

we obtain the estimate

$$|f(p_1, p_2)| \leq |G_{\nu}|_{m'} |J_1^{-\nu} f(\cdot, p_2)|_{L_m(R^{n_1})}$$

Substituting this inequality into (IV.2) we get

$$|\Pi(r) f|_{L_m(\Omega)} \leq 4\pi r^{-\frac{n_2}{2m}} |G_{\nu}|_{m'} |J_1^{-\nu} f|_{zL_m(R^n)} \tag{IV.3}$$

for $n_1 \geq 2$.

Now we proceed to the proof of the estimate (IV.1). In this time we use the representation

$$f(p) = \int G_{\delta}(p-q)(J^{-\delta} f)(q) dq \tag{IV.4}$$

and the easily obtainable estimate

$$|G_{\delta}(h(\ell + x)) - G_{\delta}(h\ell)| \leq |h|^{-n+\delta} (H_{\delta}(\ell) + H_{\delta}'(\ell + x)),$$

where the positive functions $H_{\delta}, H_{\delta}' \in L_1$. Denoting $v = \sqrt{r+h} - \sqrt{r}$ we obtain

$$|(\Pi(r+h) - \Pi(r)) f| =$$

$$|\int [G_{\delta}(k + v\omega) - G_{\delta}(k)](J^{\delta} f)(\sqrt{r}\omega - k) dk| =$$

$$= |v|^n \left| \int [G_\delta(v(\ell+\omega)) - G_\delta(v\ell)](J^\delta f)(\sqrt{r}\omega - v\ell)\,d\ell \right|$$

$$\leq |v|^\delta \int H_\delta(\ell)\,|(J^\delta f)(\sqrt{r}\omega - v\ell)|\,d\ell\ +$$

$$+\ |v|^\delta \int H'_\delta(\ell)\,|(J^\delta f)(\sqrt{r+h}\omega - v\ell)|\,d\ell$$

Taking now the $L_m(\Omega)$-norm of this inequality gives

$$|(\Pi(r+h) - \Pi(r))f|_{L_m(\Omega)} \leq$$

$$\leq C|\sqrt{r+h} - \sqrt{r}|^\delta \sup_k (|(J^\delta f)(\sqrt{r}\omega + k)|_{L_m(\Omega)} +$$

$$+\ |(J^\delta f)(\sqrt{r+h}\omega + k)|_{L_m(\Omega)}) \tag{IV.5}$$

Estimate (IV.3) and the fact that $J_1^{-\nu}$ commutes with the translations yield

$$|(J^\delta f)(\sqrt{r}\omega + k)|_{L_m(\Omega)} \leq Cr^{-\frac{n_2}{2m}} |J_1^{-\nu} J^{-\delta} f|_{L_m(R^n)} \tag{IV.6}$$

Inequalities (IV.5) and (IV.6) imply desired estimate (IV.1). The lemma is proved.

Let $R^{n_1} = R^{n'} \oplus R^{n''}$, $n' + n'' = n_1$. Since

$$L_m^\alpha(R^{n_1}) \otimes L_m^\beta(R^{n_2}) \subset L_m^\alpha(R^{n'}) \otimes L_m^\beta(R^{n''} \oplus R^{n_2}) \quad \text{if} \quad \alpha \geq \beta \geq 0$$

one obtains immediately.

Corollary IV.1. The family $\Pi(r)$ is uniformly bounded from $J_1^\nu L_m(R^n)$ into $L_m(\Omega)$, $\nu > \frac{2}{m}$, and its derivative, $\Delta_r^\delta(h)\Pi(r)$, is uniformly bounded from $J^\delta J_1^\nu L_m(R^n)$ into $L_m(\Omega)$:

$$|\Delta_r^\delta(h)\Pi(r)|_{J^\ell J_1^\nu L_m(R^n) \to L_m(\Omega)} \leq Cr^{-\frac{1}{2}(\frac{n-1}{m}+\delta)} \tag{II.7}$$

SUPPLEMENT

For the reader's convenience, this supplement lists a few well-known results from the theory of embedding theorems for various classes of fractionally-differentiable functions (for details, see [34], [35]) and the theory of singular integrals (see [34], [36], [37]), which are referred to constantly throughout the paper.

We shall state these embedding theorems as inequalities (and not in terms of spaces as usual) for functions of $S(R^m)$. Throughout, f is a function in $S(R^n)$.

$$|\Delta^s(\ell)f|_{L_m(R^n)} < C\{|f|_{L_m(R^n)} + [\int |\Delta^{s'}(h)f|^m_{L_m(R^n)} \frac{dh}{|h|^n}]^{1/m}\}$$

$$\leq C_1(|f|_{L_m(R^n)} + \sup_v |\Delta^{s''}(v)f|_{L_m(R^n)}),\quad s < s' < s'' \leq 1; \tag{S.1}$$

$$|\Delta^s(h)f|_{L_m(R^n)} \leq C\,|J^{-s}f|_{L_m(R^n)}$$

$$\leq C_1(|f|_{L_m(R^n)} + \sup_\ell |\Delta^{s'}(\ell)f|_{L_m(R^n)}),\quad s < s' \leq 1, \tag{S.2}$$

where

$$(J^s f)(k) = \int e^{ikx}\,\hat{f}(x)\,(1 + |x|^2)^{-s/2}\,dx;$$

and

$$|\Delta^s(h)f|_{L_{m_1}(R^n)} < C(|f|_{L_m(R^n)} + \sup_\ell |\Delta^{s'}(\ell)f|_{L_m(R^n)}), \tag{S.3}$$

where $1 \leq m \leq m_1 \leq \infty$, $1 \geq s' > s + n(\frac{1}{m} - \frac{1}{m_1})$.

The fundamental result for regular integrals in spaces of functions on R^1 which are summable to some power.

THEOREM. The operator

$$(T_{\pm}f)(x) = \int_{-\infty}^{\infty} \frac{f(y)\,dy}{y - x \pm i0}, \quad f \in S(R^1), \tag{S.4}$$

has a bounded extension in $L_m(R^1)$, $1 < m < \infty$:

$$|T_{\pm}f|_{L_m} \leqslant C_m |f|_{L_m}, \quad 1 < m < \infty, \quad f \in L_m(R^1). \tag{S.5}$$

REFERENCES

1. A. Ya. Povzner, Mat. Sbornik 32 (74) (1953), 109, English transl. Amer. Math. Soc. Trans. (2) 60 (1967), 755.

2. T. Ikebe, Arch. Ratl. Mech. Anal. 5 (1960), 1.

3. L. D. Faddeev, Trudy Mat. Inst. Steklov 69 (1963), English transl. Israel Program for Scientific Translations, Jerusalem; Davey, New York, 1965.

4. K. Hepp. Helv. Phys. Acta 42 (1969) 425.

5. R. B. Lavine, Commun. Math. Phys. 20 (1971), 301.

6. R. J. Iorio, Jr., and M. O'Carroll. Commun. Math. Phys. 27 (1972), 137.

7. I.M. Sigal. Preprint. Inst. Teor. Fiz. Akad. Nauk Ukr. SSR-74-19R (1974). Preprint, Tel-Aviv University (1975).

8. M. Ekstein. Phys. Rev. 101 (1956), 880.

9. F. A. Berezin. Dokl. Akad. Nauk SSSR 163 (1965), 795 - Soviet Math. Dok. 6 (1965) 997.

10. A.G. Sigalov and I.M. Sigal. Teor. Mat. Fiz. 5 (1970), 73 = Theoretical and Mathematical Physics. 5 (1970).

11. E. Balslev, Reports in Math. Phys. 5 (1974), 213, 393.

12. F. A. Berezin, R. A. Minlos and L. D. Faddeev. Proc. Fourth All-Union Mathematical Congress 1961, Vol. II, 532, 1964.

13. W. D. Amrein, V. Georgescu and J. M. Jauch, Helv. Phys. Acta 44, (1971), 407.

14. T. Kato. Trans. Am. Math. Soc. 70 (1951), 195.

15. J. M. Jauch. Helv. Phys. Acta 31 (1958), 661.

16. I. M. Cook, J. Math. Phys. 36 (1957), 82.

17. M. N. Hack. Nuovo Cim. 13 (1959), 231.

18. I. M. Sigal. Dokl. Akad. Nauk SSSR 204 (1972), 795 = Soviet Math. Dokl. 13 (1972), 756.

19. I. M. Sigal, Unpublished.

20. J. Weidmann, Bull. Am. Math. Soc. 73 (1967), 452.

21. I. M. Sigal, Comm. Math. Phys. 48 (1976), 137; 155.

22. G. M. Zhislin and A. G. Sigalov, Izv. Akad. Nauk. SSSR. Ser. Mat. 20 (1965), 835, 1261.

23. G. M. Zhislin, Math. of the USSR-Izvestia 3 (1969), 559.

24. B. Simon, Annals of Math. 97 (1973), 247.

25. J. Howland, J. Math. Appl. 50 (1975), 415.

26. L. P. Horwitz and I. M. Sigal, Preprint, Tel Aviv University (1976).

27. J. M. Combes, Preprint, Marceille (1969);
F. Aguilar and J. M. Combes, Comm. Math. Phys. 22 (1971), 269;
E. Balslev and J. M. Combes, Comm. Math. Phys. 22 (1971), 280.

28. D. Babbit and E..Balslev, Comm. Math. Phys. 35 (1974), 175; J. of Funct. Anal. 18 (1975), 1.

29. D. R. Jafaev, Math. of the USSR - Sbornik 23 (1974), 535.

30. B. Simon, Math. Ann. 207 (1974), 133.

See also: E. Balslev, Arch. Ratl. Mech. Anal., 59 (1975), 343.

31. I. M. Sigal. Teor. Mat. Fiz. 10 (1972), 249 = Theoretical and Mathematical Physics, November 1972, 165.

32. S. Albeverio, W. Hunziker, W. Schneider and R. Schrader. Hilv. Phys. Acta 40 (1967), 745.

33. O. A. Yakubovskii. Trudy Mat. Inst. Steklov (1970) = Proceedings of the Steklov Institute of Math., N110 (1970).

34. E. Stein. Singular Integrals and Differential Properties of Functions, Princeton, 1970.

35. S. M. Nikolskii. Approximation of Functions of Several Variables and Imbedding Theorems, Springer-Verlag, 1975.

36. N. Dunford and J. T. Schwartz. Linear Operators, Vol. 1, New York, Academic Press, 1959.

37. A. Zygmund, Trigonometrical Series, Vol. 2, Cambridge University Press, 19.

38. I. M. Gellfand and G. E. Shilov. Generalized Functions, Vol. 1, Moscow, 1959.

39. T. Kato, Math. Annalen 162, 1966, 258.